63rd Israel Annual Conference on Aerospace Sciences 2024

Haifa, Israel
9 May 2024

ISBN: 978-1-7138-9694-4

Printed from e-media with permission by:

Curran Associates, Inc.
57 Morehouse Lane
Red Hook, NY 12571

Some format issues inherent in the e-media version may also appear in this print version.

Printed by Curran Associates, Inc. (2024)

For permission requests, please contact Faculty of Aerospace Engineering, Technion - Israel Institute of Technology at the address below.

Faculty of Aerospace Engineering
Technion - Israel Institute of Technology
Department of Aerospace Technology
Haifa 32000
Israel

Phone: (04) 829-2260
Fax: (04) 829-2030

aerospace.technion.ac.il/

Additional copies of this publication are available from:

Curran Associates, Inc.
57 Morehouse Lane
Red Hook, NY 12571 USA
Phone: 845-758-0400
Fax: 845-758-2633
Email: curran@proceedings.com
Web: www.proceedings.com

TABLE OF CONTENTS

AERODYNAMICS & AEROACOUSTICS

AERODYNAMICS, HYDRODYNAMICS AND AEROACOUSTICS

AEROELASTICITY & FLUID-STRUCTURE INTERACTION

AEROSPACE DESIGN AND MANUFACTURING

AEROSPACE DESIGN, MANUFACTURING AND MULTI-DISCIPLINARY OPTIMIZATION

ASTRODYNAMICS AND SPACE SYSTEMS

DATA, AI AND AUTONOMY

FLOW CONTROL AND CFD

FLUID MECHANICS

GUIDANCE, NAVIGATION AND CONTROL I

GUIDANCE, NAVIGATION AND CONTROL II

PROPULSION AND COMBUSTION

SOLID MECHANICS

On the Performance and Longitudinal Stability of Custer Channel Wing (CCW)

By

Moshe Zilberman, Sharon Gordani

Azrieli College of Engineering, Jerusalem

Custer's channel wing aircraft was never certified by the FAA. After 40 years of a lifetime project, Willard Custer gave up in 1970 and his CCW-5 last version of a CCW aircraft became a full-size exhibit in the aviation museums. The CCW concept was never forgotten and was examined by some leading aviation experts to facilitate the concept for innovative low-cost STOL air vehicles. To mention some, The US Army Air Force and later NACA Langley were thoroughly experimenting with a full-size model in their full-size wind tunnels in the 50s and 60s of the previous century. Later on, between the 80s and the year 2000 the team at Georgia Tech was extending the concept to include air jets that increase the circulation around the wings. The idea of incorporating the CCW on innovative airliner designs with over-the-wing engines is recommended by NASA and Boeing experts.

The motivation for examining this configuration at Azrieli College is to assess the feasibility of retrofitting UAVs and short-haul airliners for use on short airfields in rural and urban environments. In a step-by-step procedure, we investigated the concept's advantages and weaknesses in the last six years as mentioned in the earlier works.

The present investigation follows a demo flight of an R/C DHC6 1:17 model aircraft, retrofitted with a CCW and reduced the take-off run to 30% of its ground run with a straight wing. We present here the results of aerodynamic coefficients, obtained for the 3:1 scale relative to the previous model with a generic body and FX63-137 wing. The measurements confirmed the increase in the lift coefficient for a wing-body (WB) configuration from 1.5 at 15^0 (the stall angle for the straight wing) to the lift coefficients, for the CCW model, between 2.8 to 4.7 for thrust coefficients between 1.7 to 12, respectively. Similar results were obtained for the wing-body-tail (WBT) configuration with a lift coefficient of 1.75 at 15^0 for the straight wing and lift coefficients between 3.0 to 4.9, for the CCW model and the same range of thrust coefficients. It is demonstrated that trim angle and positive static stability are achievable, in contrast to old pessimistic experiments by NASA Langley. Another original finding in the present work is that at low negative angles of attack the $C_L^{1.5}/C_D$ and C_L/C_D ratios for propeller-driven aircraft maximum endurance and range, respectively, can be regained due to the positive component of the induced lift from the channel. The conclusions obtained for the present investigation and previous ones are that retrofitting UAVs and short-haul airliners, especially twin-engine, is very attractive due to significant reduction in ground run and the moderate changes and cost. To examine the retrofit of an existing vehicle or a fresh design of a CCW aircraft, further analysis is required to quantify the parameters of the retrofit, like channel size and motor thrust coefficients.

Integrating ultrasonic and photonic forces.

Ariel Sharon, Levy Yeshayahou.
Faculty of Aerospace Engineering, Technion – Israel.

ABSTRACT.

In the 1970s, Arthur Ashkin's groundbreaking experiments revealed an intriguing aspect of light: its ability to carry momentum. By focusing a laser beam, Ashkin managed to suspend a dielectric particle within its divergent region, defying gravity through the intricate interaction between the laser's electromagnetic radiation and the suspended particle. This achievement marked a pivotal moment in the field of microparticle manipulation, demonstrating the potential of light as a tool for precise control at the microscale. Following Ashkin's pioneering work, subsequent studies have delved deeper into the multifaceted relationship between light and particles. Researchers uncovered a rich tapestry of phenomena, from optical levitation and droplet ejection to laser-induced cavitation, optical trapping, droplet fragmentation, and deformation. Each of these phenomena offers unique insights into the complex interplay between light and matter, opening up new avenues for exploration and technological innovation.

In parallel to these advancements in optical manipulation, an alternative method for particle manipulation has emerged: acoustic levitation. By subjecting objects to a sound field in a fluid medium, researchers can generate a consistent time-averaged force known as acoustic radiation force, allowing for the stable suspension of objects against gravity without physical contact. This approach offers a compelling alternative to traditional optical methods, particularly for manipulating larger objects on a macroscopic scale.

Our research endeavors aim to explore the synergistic effects arising from the combined forces of acoustics and photonics. While optical tweezers excel at manipulating small particles within a limited size range, acoustic levitation offers the advantage of producing stronger force trapping on a larger spatial scale, enabling the manipulation of millimeter-sized drops. However, the drawback of acoustic levitation lies in its inability to offer precise control over the levitated subject, which we seek to address through our research efforts of combining acoustic and photonic forces.

Initial experiments have shown promising results, demonstrating the lateral manipulation of levitated water droplets with diameters ranging from 80 to 150 micrometers. We have achieved controlled acceleration of these droplets at a rate of 0.2 mm/s^2 over a distance of 170 micrometers in under 850 milliseconds. These outcomes represent the effects of acoustic damping force, radiation pressure, drag, and gravity. These acoustic damping force pertains to the forces applied to the droplet within a static acoustic environment. These forces function as a mechanism for damping, aimed at diminishing the motion of the droplet. Our experiments consistently yielded measurable results, establishing a robust basis for the practicality and effectiveness of macro droplet manipulation thereby divulging a plethora of new design opportunities for macro-droplet dynamics study.

Towards Low Distortion Airborne Optical Turrets: Leveraging Computational Aero-Optics and Data-Science Methods

Michael Weidenfeld and Ori Haber

Aeroballistics Dep., Land Division, Elbit, POB 1044 Ramat Hasharon 4711001, Israel
michael.weidenfeld@elbitsystems.com

Aero-optic design of low-distortion airborne optical turrets, presents multiple challenges, among which are the development of a high-fidelity computational framework, followed by conceptual design exploration, simulation of turret interaction with aircraft and the aeroacoustics of nearfield sources, and seeking advanced algorithms for signal error reduction. The present work details efforts to address these challenges, some of the insights gained regarding the designs examined, and the adaptive optics approach leveraged to remove most of the phase distortion and produce a high performance optical system.

Our approach is structured around a four-step computational framework, bridging between the aero- and the optics. This framework is comprised of four key computational steps: 1) executing high-fidelity aerodynamic flow computations to evaluate the density and optical index of refraction fields; 2) interpolating data from the CFD mesh to an optic mesh representing wavefront propagation directions; 3) harnessing paraxiality to apply ray tracing on the optic box mesh to determine the aberrated wavefront (referred to as the OPD - Optical Path Difference); and 4) using computational Fourier optics to propagate the distorted signal to the far-field, focusing on aero-optic effects by excluding atmospheric turbulence and other wavefront errors, to isolate and assess the system's optical performance.

In our exploration of turret designs, we focused on classical spherical turrets, featuring both flat and conformal windows, with a particular emphasis on forward-looking orientations. This initial choice sought to navigate away from the complexities of wavefront propagation through turbulent wakes and shear layers. However, our subsequent results showed that looking forward through the flow stagnation zone predominantly resulted in defocus-type aberration—a finding in line with theoretical expectations for an ideal stagnation flow and isentropic relations. This discovery prompted a shift to flat window designs, akin to "wedge" flows, which altered the primary aberration from defocus to more manageable tip/tilt aberrations, thus offering a potential solution to the challenges identified with blunt spherical turrets.

Another aspect of our research was the simulation of the turret's interaction with the aircraft and the aeroacoustics of nearby stores and sound generating sources. These simulations shed light on the role of real-world turret environment and its impact on optical performance. The results currently exhibit high order spatial structures superimposed on very energetic low order modes, thus pointing towards better understanding the nature of the spatio-temporal wavefront signal via data science methods.

The resulted wavefronts were then decomposed using Dynamic Mode Decomposition (DMD) and Proper Orthogonal Decomposition (POD). The most notable observation was obtained via POD, showing across all considered viewing angles and flight conditions, that there exists a dominant spatial mode unique to each scenario, capturing almost the entirety of the signal's variance or energy. Remarkably, the singular value of this mode is two orders of magnitude higher than any subsequent ones. The existence of such a singular mode capturing most of the signal's variance calls for a rather straightforward error mitigation method. By employing a deformable mirror, we can correct the majority of phase errors in a consistent manner, avoiding the need for continuous broad-spectrum adjustments.

Our findings encourage a passive aero-optic design, steering clear of active flow controls. This approach, reflects a strategic decision to focus on design optimizations that can be practically implemented without the complexities and potential limitations associated with active flow control methods. Instead, shifting from spherical to flat window configurations not only facilitates a passive design but also significantly leverages adaptive optics for error correction, challenging the conventional skepticism towards using adaptive optics for airborne turret aberrations.

In conclusion, our work presents a comprehensive and methodical computational approach, informed by insights from data science techniques, to the aero-optic design of low distortion airborne optical turrets, showcasing the potential of adaptive optics in overcoming turbulent induced aberrations in passive flat window designs for enhancing optical performance.

Experimental Study of the Side-Edge Vortex System on a Supercritical Wing Model

H. Ben-Gida *
Technion – Israel Institute of Technology, Haifa 3200003, Israel

S. Baba†
University of Toronto Institute for Aerospace Studies, Toronto, ON, M3H 5T6, Canada

Oksana Stalnov ‡
Technion – Israel Institute of Technology, Haifa 3200003, Israel

Stéphane Moreau§
Université de Sherbrooke, Sherbrooke, Québec J1K 2R1, Canada

P. Lavoie¶
University of Toronto Institute for Aerospace Studies, Toronto, ON, M3H 5T6, Canada

This study presents an experimental characterization of the vortex system generated on the side edge of a canonical cantilever wing with a supercritical profile. Experiments were conducted in the newly commissioned hybrid anechoic wind tunnel (HAWT) at the University of Toronto. The HAWT test section was configured with either Kevlar side walls for acoustic measurements or glass walls for flow visualization measurements. Static pressure taps, remote surface microphones, and phased microphone array measurements were utilized to characterize the model's aerodynamics and noise spectra. Stereoscopic particle image velocimetry (PIV) measurements were performed to characterize the flow field around the side edge, focusing on the dynamics of the side-edge vortical structures as they develop downstream. Measurements were performed at a geometric angle of attack of 10° and at a freestream velocity of $U_\infty = 50\,\mathrm{m/s}$, corresponding to a Mach number of $M_\infty = 0.15$ and a chord-based Reynolds number of $Re = 1 \times 10^6$. Results revealed a triple-vortex system (TVS) to develop along the side edge, rotating in a helical manner as it travels downstream. Proper orthogonal decomposition (POD) and vortex wandering analysis shed light on the dynamics of the TVS and its potential noise-generating flow mechanisms as it evolves along the side edge. Coherence analysis between the surface pressure fluctuation measurements across the side edge displayed a dominant hump at $f \sim 1,300$ Hz, which is associated with the primary vortex. Coherence analysis between surface pressure spectra and far-field microphones shows a dominant hump at a similar frequency. Furthermore, integration of the source localization maps (obtained by phased microphone array measurements) around the side edge reveals a prominent hump in the spectra around the same frequency, thus confirming the roll-up of the primary vortex from the side edge to the suction surface as the dominant noise source. The results presented herein provide insight into the noise-generating flow mechanisms of a side edge with a supercritical airfoil having a highly cambered aft section. Such knowledge may potentially lead to the development of improved noise reduction strategies for aircraft flap side edges, thus diminishing the impact of airframe noise on communities surrounding airports.

*Research Fellow, Faculty of Aerospace Engineering, bengida@technion.ac.il
†PhD student, Institute for Aerospace Studies.
‡Associate Professor, Faculty of Aerospace Engineering.
§Professor, Université de Sherbrooke Department of Mechanical Engineering.
¶Professor, Institute for Aerospace Studies.

Aeroacoustic Characterization of a High-Lift Multi-Element Configuration

H. Ben-Gida *
Technion – Israel Institute of Technology, Haifa 3200003, Israel

M. K. Okoronkwo†, D. G. Geneau‡
University of Toronto Institute for Aerospace Studies, Toronto, ON, M3H 5T6, Canada

Stéphane Moreau§
Université de Sherbrooke, Sherbrooke, Québec J1K 2R1, Canada

Oksana Stalnov ¶
Technion – Israel Institute of Technology, Haifa 3200003, Israel

P. Lavoie‖
University of Toronto Institute for Aerospace Studies, Toronto, ON, M3H 5T6, Canada

This study presents the aerodynamic and aeroacoustic characteristics of the two- and three-element configurations of the canonical high-lift 30P30N airfoil. Experiments were conducted in the newly commissioned hybrid anechoic wind tunnel (HAWT) at the University of Toronto, where the HAWT test section is configured with Kevlar side walls to allow for simultaneous aerodynamic and acoustic measurements. Static pressure taps, remote surface microphones, and phased microphone array measurements were utilized to characterize the aerodynamics and noise spectra of both configurations. Measurements were performed at geometric angles of attack ranging from 0° to 15° at a freestream velocity of $U_\infty = 50$ m/s, corresponding to a Mach number of $M_\infty = 0.15$ and a chord-based Reynolds number of $Re = 1.43 - 1.5 \times 10^6$. The experimental results are compared with unsteady Reynolds-Averaged Navier-Stokes (uRANS) simulations to gain better physical insights into the flow phenomena associated with the noise sources in multi-element configurations. The results of the mean pressure coefficient distribution are shown to be in good agreement with the uRANS simulations, as well as with measurements of similar models in other HAWTs. Generally, the mean surface loading in the flap region is insensitive to variations in the angle of attack, suggesting minimal variation in broadband noise with loading. Surface pressure fluctuations measured in selected regions (slat, main element, and flap), as well as far-field noise measurements and source localization maps, reveal distinctive patterns and provide insight into the noise-generating mechanisms of the 30P30N airfoil. Tonal peaks are found both in the flap cove and on the flap, matching the tonal noise produced by the slat when in the three-element configuration. The tonal peaks in the flap region are shown to be induced by acoustic waves traveling downstream from the slat cove. These acoustic waves interact with the flow within the flap cove to generate upstream traveling acoustic waves, thus forming an acoustic feedback loop between the slat and flap coves that further amplifies the narrowband tonal peaks in both coves. These tones disappear from the flap region when the slat is removed in a two-element configuration, while the broadband noise level remains unchanged. The results from this study advance our knowledge of noise generation mechanisms in high-lift multi-element configurations and set the basis for developing more accurate physics-based noise models to improve the acoustics of future airframe designs.

*Research Fellow, Faculty of Aerospace Engineering, bengida@technion.ac.il
†PhD student, Institute for Aerospace Studies.
‡MSc student, Institute for Aerospace Studies.
§Professor, Université de Sherbrooke Department of Mechanical Engineering.
¶Associate Professor, Faculty of Aerospace Engineering.
‖Professor, Institute for Aerospace Studies.

Title: Comparison of Traditional Acoustics and Psychoacoustics Analysis of a Rotor in Hover
Authors: Aleksandra Kvurt, Dr. Aharon Karon, Shai Alexanderoni, and Danny Abramov

Unmanned Air Vehicles (UAVs) and Urban Air Mobility craft have many applications in the civil sector. However, the noise produced by these platforms combined with their proximity prevents their acceptance by the public. This means, that in addition to the essential design considerations of performance, aerodynamics, and traditional acoustics, sound perception (psychoacoustics) should be incorporated. Psychoacoustic analysis has been validated through listening experiments, where a representative group of people rate their perception and annoyance with different sounds.

During this study, the far-field noise of an off-the-shelf rotor in hover conditions was measured and analyzed using both conventional and psychoacoustic techniques. To demonstrate the capabilities of the psychoacoustic analysis, the Sound Quality Metrics such as loudness, sharpness, roughness, fluctuation strength, and annoyance of the rotor noise were derived by processing the experimental data. A key aspect of these tests involved manipulating the rotational per minute (RPM) control signal to replicate real-world flight dynamics. Thus, three scenarios were investigated: (1) a constant RPM command, (2) an RPM profile mimicking a real UAV maintaining hovering, and (3) an RPM profile mimicking a real UAV performing a maneuver. Classical and psychoacoustic analysis techniques were then employed to analyze the effect on perceived noise under these conditions.

Initial results using the classical Fourier analysis show that the blade pass frequency (BPF) is the most significant feature of the noise for all three cases. Additionally, there is virtually no difference between the constant RPM and hover control signal cases. However, the maneuver case generates the most noise overall. This is because fluctuating rotor speed during maneuvers produces more noise energy, even though it's spread across a wider range of frequencies. These results are illustrated in Figure 1. It should be observed that all three cases generated noise in the same level range.

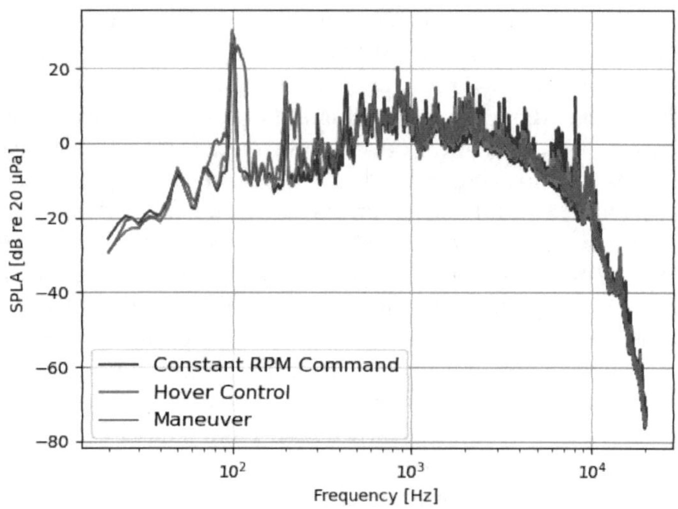

Figure 1: Average power spectra of the three control cases. ϑ= 48°, Δf = 2.5 Hz. The noise was propagated from the altitude of 100 m with atmospheric attenuation taken into account.

Initial results from the psychoacoustic analysis are shown in Figure 2. As seen in Figure 2, using the ISO 532-3 loudness model, the most impactful frequencies are between 2 kHz and 7 kHz, and not the BPF. Additionally, the transient nature of the control signal commands increases the loudness significantly.

We observed a clear trend of increasing loudness simply by perceiving the signals, and this aligns with the findings of psychoacoustic analysis. This suggests that humans are sensitive to frequency changes that may contribute to the increased annoyance caused by UAVs, and psychoacoustic analysis is capable of detecting this where traditional analysis may not.

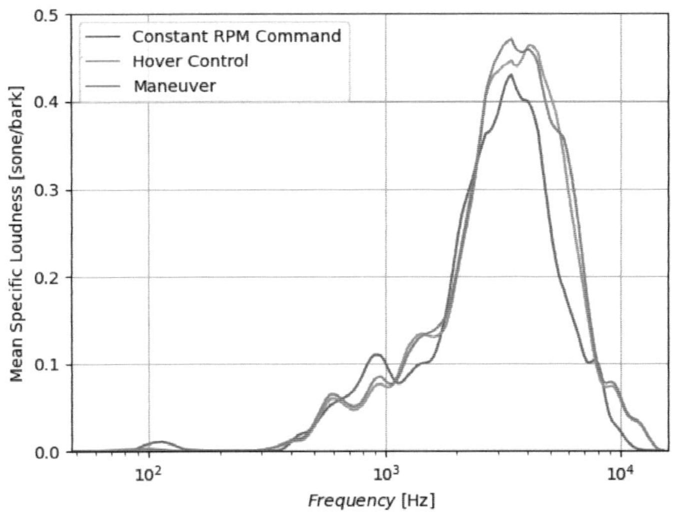

Figure 2: Mean over time specific loudness of the three control cases. ϑ= 48°, frequencies are with respect to the Bark scale. The noise was propagated from the altitude of 100 m with atmospheric attenuation taken into account

Classifying structure and air borne noise pathways in urban structures subjected to subsonic wind flow

Saar Golan[1*], Ofer Avishai[1], Rami Shahar[1], Yuval Wulkan[1], Egor Dyunin[1], Yair Gur Lavi[1], Yoram Rapoport[1], Anatoly Kovadlo[1]

We introduce a novel approach for classifying aeroacoustic and vibroacoustic pathways in urban structures. Conjugated wind tunnel, vibrational and acoustic experimental characterization of the structure and Computational Fluid Dynamics (CFD) / Finite Element Analysis (FEA) / Fluid Structure Interaction (FSI) simulations are used to generate frequency dependent vibroaeroacoustic heat maps. This representation portrays the structure acoustic signature and pinpoints specific air vs. structure borne noise pathways bands. The test subject is the Weizmann Institute new chemistry facility. The building will house a wide range of advanced scientific instrumentation that is highly sensitive to noise and vibration and must comply with unique vibration criteria. Its envelope contains airfoil-like shading elements (slabs, Fig. 1) that are amenable to wind-induced vibrations and can transfer kinetic energy to the main structure and induce acoustic perturbations in regions close to the shell. CFD, FEA and FSI simulations and vibration (Hammer) experiments were conducted at Dynamica Design facilities (Fig. 2). Wind tunnel testing of a representative test unit (~800 Kg) mounted with multiple accelerometers and microphones was conducted at the Israel Aircraft Industry (IAI) subsonic wind tunnel (Fig. 1). Statistical testing correlated specific accelerometers and microphones (e.g., Pearson correlation & Mann-Whitney, Fig. 2). Results confirm specific acoustic/vibration pathways (e.g., drumming or rigid body vs. membranal modes). Vibroaeroacoustic heat maps identify structural components amenable to flow vs. vibration induced pathways and their signature patterns (Fig. 3). But more important, they could facilitate a standardized approach for acoustic background levels characterization in urban settings and machine learning based identification of subsonic airborne anomalies (e.g., drone & unmanned aircraft).

[1]Dynamica Design, *Corresponding author (Saar@dynamica.net)

Keywords - Aerodynamics, Fluid Dynamics, Aeroacoustics

Figure 1: Computational and experimental models (a) *Architectural rendering of the Weizmann Institute new chemistry facility envelope [1]. The shell comprises ~6500 slabs, each ~1.5 [m] length, to a total of ~10 km. The potential to induce vibrations and noise is evident.* ***(b)*** *The test unit design for the study comprises three slabs. The central slab is the one representative of a repetitive pattern conditions.* ***(c)*** *The experimental prototype (~800 Kg). Selected microphones and accelerometers are indicated by the red arrows.*

[1] https://mochly-eldar.com/project/weizmann-chemistry/

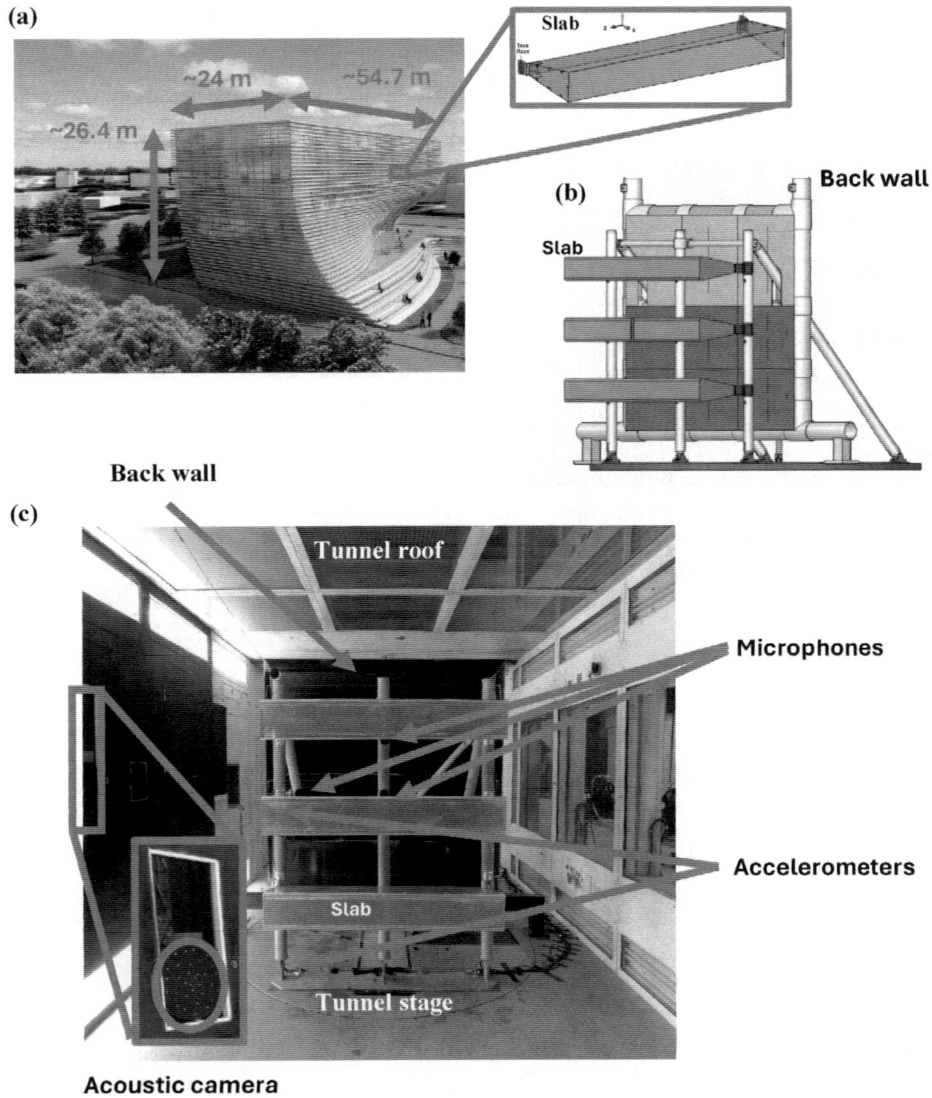

Figure 2: CFD and FEA simulations (a) *CFD analysis of the building cross section. Color bar on the right and arrow size indicate speed (m/s). Arrow direction indicates velocity. The analysis recaps the well-established front, back and rooftop building eddies (Comsol). (b) FEA derived natural frequencies (Abaqus). Fundamental rigid bogy modes are received for ~20 Hz. Membranal modes are present at ~50-60 Hz.*

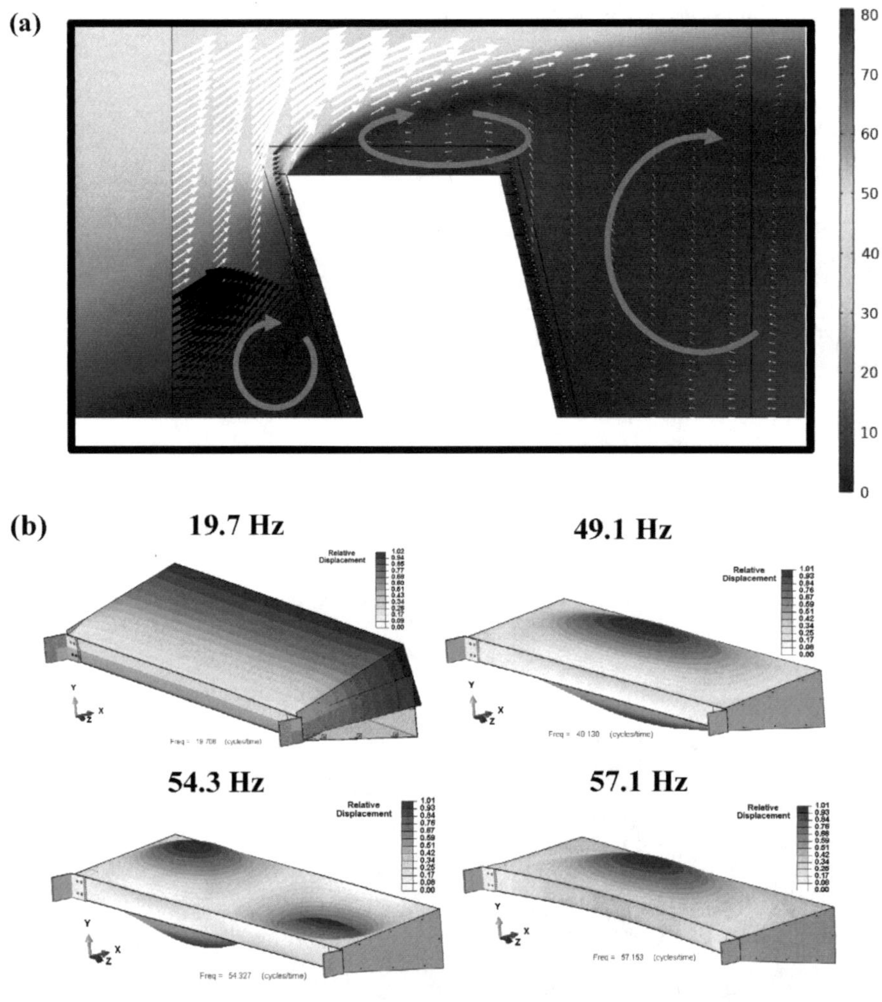

Figure 3: Experimental results (a) *Slab internal acoustic resonance (drumming effect). Noise generated can be significant (~10 dB above environment). Mitigation – side caps covering.* ***(b)*** *Incident wind speed considerably affects slab accelerations.* ***(c)*** *Pressure sensor mounting scheme for the slabs.* ***(d)*** *Pearson's correlation coefficients heat map for the various pressure sensors. Noise potential is implied by low pressure correlations indicating high spatial gradients (flows), particularly for adjacent sensors.* ***(e)*** *Scatter plots of the pressure correlations. High slot - environmental pressure correlation (P13 between slabs vs. P10-12 outside) indicates minimal whistling potential.*

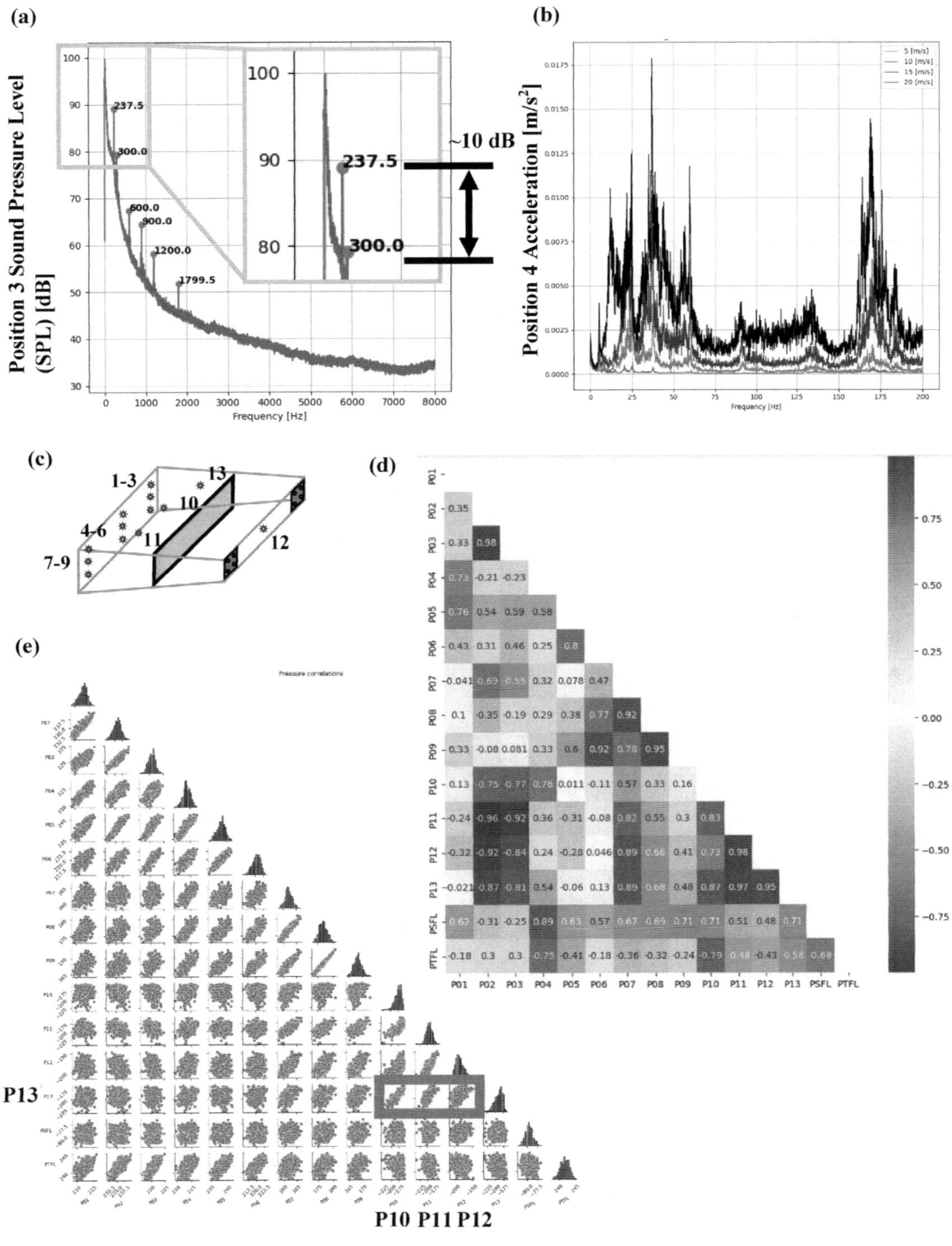

Topology Based Multi-Fluid Flow Model

Ido Silverman*
Soreq – NRC, Israel

This work presents a new multi-fluid flow model that accounts for the actual flow topology (dispersed or separated) and provides a unified model for all flow regimens. The basis of the new model is a physical insight that guides the averaging process of the multi-fluid conservation equations, which cannot be purely a mathematical treatment. Physically correct averaging processes have been achieved by going back to the control volume model and considering the contributions of each phase on the conserved variable being studied (i.e., moving from single phase averaging process to the whole mixture). The study defines a topology function $\zeta_1(\alpha, \nabla\alpha, r, x, t)$ that equals 1 if phase 1 is fully continuous in a region of radii r around point x at time t. The function equals 0 if phase 1 is fully dispersed in phase 2 (i.e., phase 1 is divided to several macroscopic particles which are connected only through phase 2); and any number in between for intermediate cases. The new topology function is used to divide the mixture averaged conservation equations into phasic equations in a way that correctly accounts for transfer of stresses, heat, etc. in the mixture. It is shown that from this perspective, a unified model that is able to correctly describe the interactions between the phases in the flow field, starting from dispersed conditions and moving to separated flows, can be provided.

Two variants of the model, for two flow regimes have been further developed and studied for their mathematical and numerical properties. One is a fully non-equilibrium model with different pressures, temperatures and velocities for each phase while the other assumes pressure equilibrium. The first model assumes two separated fluids flowing side by side and presents different wave speeds for each phase that depend on the phase properties. The other assumes a flow where one phase in finely dispersed in the other and presents only one wave speed for the mixture that are dependent on the properties of the continuous phase and relative velocity of the dispersed phase. The combined model is applicable for the full range of void fraction of zero to one ($0 \leq \alpha \leq 1$) and thus cover all possible multiphase scenarios.

The new model has been implemented in a compressible multi-phase flow solver using the HLLC flow solver. Some 2D compressible multiphase flow benchmarks have been simulated to demonstrate the capability of the model and study the effect of the topology functions model on the solution. It has been demonstrated that the assumed flow topology influence the structure of the flow; mostly, when velocity and thermal none-equilibrium conditions in the mixture are accounted for.

*Email: ido@soreq.gov.il

DEPARTMENT OF
AEROSPACE ENGINEERING

FMLAB

Computer Vision Algorithms Application for Tracking and Characterization of Non-Spherical Bubbles in Cavitating Flows

Samuel Gabison*
Advisor: Dr. Igal Gluzman†

Technion - Israel Institute of Technology, Haifa 3200003, Israel

Bubble dynamics effects play a significant role in the behavior of cavitating flows. Many studies and efforts have been devoted to understanding bubble dynamics, focusing on the characterization of different aspects of cavitation and bubble behavior in different flow regimes, where the complex coupled interactions between nonspherical bubbles in the cavitating clouds remain an extremely challenging task for experimental measurements. In [1], computer vision (CV) algorithms have been employed to non-intrusively extract quantitative data from high-speed imaging of the bubble spatial–temporal evolution, breakup dynamics, and cavitation inception mechanisms in aviation fuels flowing in a stable liquid state inside the diverging portion of a CD nozzle. In this work, we reproduced these algorithms and further extended them with new bubble tracking capabilities by adapting Kalman filter-based existing CV functions from MATLAB Computer Vision Toolbox. The novel procedure was first tested and validated in a controlled set of toy problems that simulated various scenarios of bubble kinematic behavior, including tracking bubble rapid growth and collapse that was simulated via the Rayleigh–Plesset equation. Then, we utilized our tracking code on existing experimental data from [1]. Our extended CV procedure allows us to obtain unprecedented new quantitative data on non-spherical bubble kinematics that could not be obtained with the CV code that was used in [1]; these include monitoring bubble size evolution along Lagrangian paths of each individual bubble after the breakup. These results shed some light on the kinematic evolution of a group of nonspherical cavities and have great importance in characterizing and modeling complex bubble dynamics mechanisms in cavitating flows.

References

[1] I. Gluzman and F. O. Thomas, "Characterization of bubble dynamics in the nozzle flow of aviation fuels via computer vision tools," Int. J. Multiph. Flow, p. 104133, 2022.

*Undergraduate Student, Faculty of Aerospace Engineering. E-mail: gabison@campus.technion.ac.il
†Assistant Professor, Faculty of Aerospace Engineering. E-mail: igal.gluzman@technion.ac.il

Computer vision algorithms application for the characterization of bubbly shock-wave morphology and their coupled interactions with bubbles in aerated cavitating flow

Elad Zur*, Igal Gluzman**

Technion -- Israel Institute of Technology, Haifa 32000, Israel

Cavitation bubble formation and collapse have high importance in developing engineering systems involving internal or external liquid flows, such as pumps and hydrofoils. The collapse of large cavitation cloud structures within internal flows can generate bubbly shocks, which may cause additional damage to the system. The complex coupled interactions between nonspherical bubbles in the cavitating cloud and bubbly shocks are extremely challenging to obtain from experimental measurements. It requires the simultaneous monitoring of spatial-temporal evolution shock waves in the cavitation cloud with monitored bubble dynamics. In this study, we have reproduced and expanded the Computer Vision (CV)-based data processing code of [Gluzman and Thomas, IJMF 2022] for bubble detection with new shock-wave detection capabilities from high-speed imaging recordings to obtain valuable data on shock wave morphology evolution and their coupled interactions with nonspherical cavity voids in cavitating flow. For that task, we have utilized the enhanced gradient shadowgraphy technique proposed by [Gluzman and Thomas, PRF 2022] to detect the presence of shock waves in aerated cavitating flows, which we couple with our CV code blob analysis procedures to detect and characterize the spatial-temporal evolution of shock waves simultaneously with bubbles. We first compare and validate our reproduced detection code with experimental results from [Gluzman and Thomas, IJMF 2022] to characterize bubble breakup kinematics in CD nozzle flow. Then, we validate our method to obtain shock speeds with results reported by [Gluzman and Thomas, PRF 2022] in aerated cavitating flows in CD-nozzle, and then use our code to obtain new quantitative data on shock-wave morphology and coupled interactions with injected bubbles, which understanding is of high importance for the development of new models for predicting the internal flow cavitation physics in different liquids.

*Dept. of Aerospace Engineering. eladzur@campus.technion.ac.il
** Asst. Prof., Dept. of Aerospace Engineering. igal.gluzman@technion.ac.il

Modified Pattern Free-Surface Synthetic Schlieren for Non-Uniformly Strained Liquid Surface

Hillel Mermelstein* and Yuval Dagan†

This research expands upon the free-surface synthetic Schlieren technique[1], which is used to measure the surface height-map of the free-surface of a liquid. The aforementioned method, which utilizes the image refraction of a random dot pattern through a free-surface to determine the surface gradients, is adapted in this study to a case where the pattern may take some arbitrary three-dimensional shape. This is shown to be theoretically valid under a small pattern slope approximation. Our method is then verified against the previously mentioned, flat-pattern, free-surface synthetic Schlieren technique by resolving the free-surface heights of plane waves across a channel, showing similar results. An example of such a setup, shown in figure 1a, includes a rectangular channel containing a transparent liquid with a random dot pattern placed on top of a cosine-shaped profile. The reconstructed fluid height presented in figure 1b shows good agreement with similar waves measured using the classical technique with a flat pattern. This new method may increase the resolution in low-amplitude regions by increasing the surface-pattern distance below these regions and correspondingly reduces the sensitivity in high-strain regions by decreasing the surface-pattern distance. Geometries are investigated in cases where surface waves have regions of very high strains followed by regions of very low amplitudes, being able to resolve the wave in both regions simultaneously. This shows promise in resolving multi-scale surface waves in highly viscous liquids or waves travelling over barriers, which may include very high amplitude regions quickly followed by very low amplitude regions.

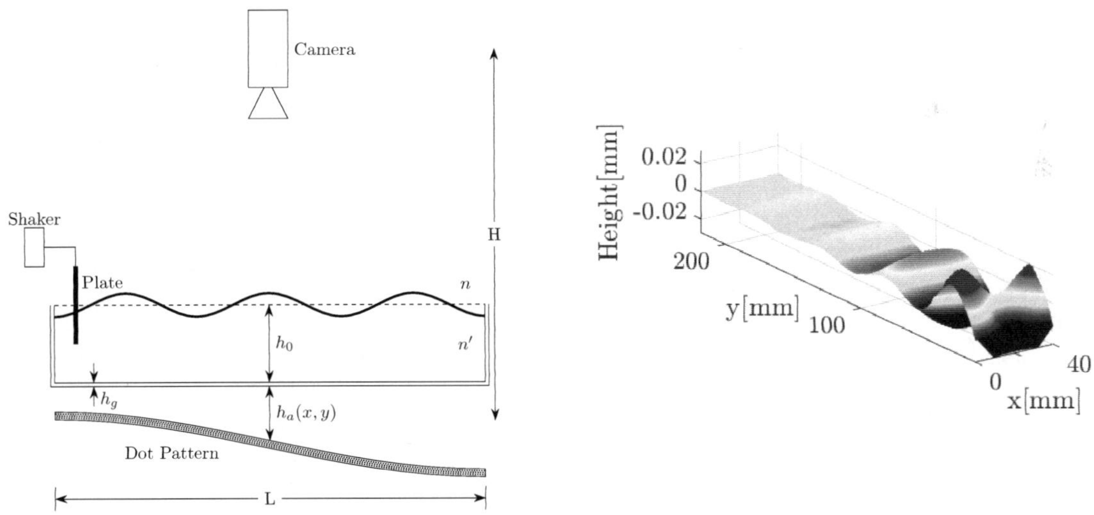

(a) (b)

Figure 1: The experimental setup shown in (a) includes a random dot pattern placed on a cosine-shaped profile at a distance H from the camera. The fluid is perturbed by a shaker vertically oscillating a submerged plate. This produces the reconstructed plane wave shown in (b). The saddle shape at the beginning of the channel is caused by the edges of the wave-maker plate inducing waves in the perpendicular direction.

*Master's Student, Faculty of Aerospace Engineering, Technion - Israel Institute of Technology
†Assistant Professor, Faculty of Aerospace Engineering, Technion - Israel Institute of Technology
[1]Moisy et al., *Exp Fluids* **46** (2009).

Development of a Thermoelectric Energy Harvesting System Integrated with a Phase Change Material Heat Sink

Daniel Jalontzki[1], Moshe Bukai[2], and Yoram Kozak[1]

[1]*Tel Aviv University, School of Mechanical Engineering, Tel Aviv, 69978, Israel*
[2]*Soreq Nuclear Research Center (SNRC), Yavne 81800, Israel*

Reliable power generation for autonomous systems without any human contact, e.g., in outer space, is an outstanding engineering challenge. Energy harvesting systems based on a thermoelectric generator (TEG) integrated with a phase change material (PCM) heat sink are a possible solution to this problem. For instance, these systems can utilize intermittent temperature differences or heat sources for remote power generation in spacecrafts. The hot side of the TEG is in contact with a heat source, whereas the cold side of the TEG is attached to a finned PCM heat sink. Heat is conducted from the fins to the PCM, which melts and absorbs the heat. Thus, a relatively constant temperature difference can be maintained across the TEG for the heating period. Although this type of systems was explored in the past, simple design guides for an optimal heat sink configuration that achieves a maximum electrical energy output for given conditions are still unknown. In the current study, we carry out detailed 2-D and 3-D numerical simulations that predict the transient heat transfer and phase change processes associated with this special type of energy harvesting system. We demonstrate that for a given fin array, there is an optimal heat flux that leads to a maximum electrical energy output. According to these detailed simulation results, we develop a simplified quasi-steady model that provides quick estimations of the optimal heat flux for given system conditions. Also, according to our optimization results, a new experimental setup is designed for extensive validation of our theoretical predictions and further investigation of the system performance. The current research outcomes will assist in the development of improved energy harvesting systems for a variety of aerospace applications.

IACAS

Investigation of Lift Coefficient Models and Unsteady Phenomenon During Pitching Motion

Naama Aliskevicius[*1] **and Oksana Stalnov**[2]

[1,2]Faculty of Aerospace Engineering, Technion — Israel Institute of Technology

*contact: naama.ali@campus.technion.ac.il

The aerodynamic response due to an airfoil in motion has generated widespread attention for nearly a century. This interest in the topic is motivated by an engineering requirement for an accurate description of unsteady aerodynamic loads, as it is essential in separated flows, noise prediction, and flutter. Many modern problems involve the evaluation of structural nonlinearities that necessitate using time-domain aerodynamic models.

Aerodynamic loads in unsteady flows can be hard to predict and therefore model, particularly at high angles of attack when the boundary layer separates. These aerodynamic loads affect the efficiency of an aircraft and specifically the lift generated by the wings. This study explores and improves lift coefficient models in low-Reynolds number flows over airfoils during pitching motion, for both attached and separated flows.

We experimentally investigate instantaneous pressure fluctuation on a symmetric NACA 0018 airfoil, at Reynolds numbers of $1.5 \times 10^5 - 3 \times 10^5$, under periodic pitching motion, as seen in figure 1. Miniature pressure sensors were embedded inside the airfoil to measure instantaneous pressures and the corresponding aerodynamic loads (Lesovoy *et al.*, 2024).

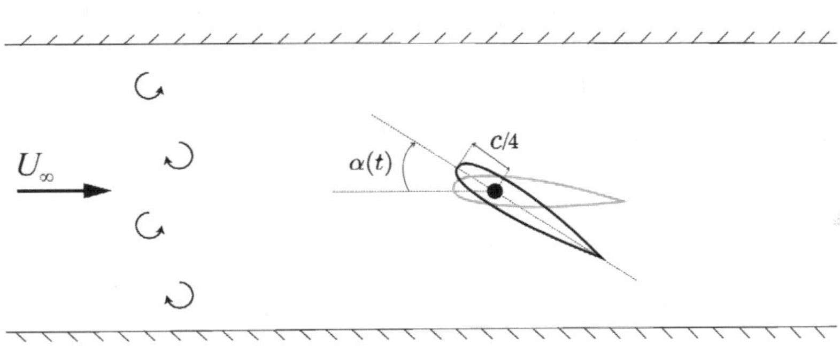

Figure 1: Top view of the experimental set-up. A two-dimensional NACA 0018 airfoil oscillates about the quarter-chord axis in grid-generated turbulence (Lesovoy *et al.*, 2024).

Our primary objective is to provide a model that can predict with sufficient accuracy the aerodynamic loads during motion in the time domain. We commence with the classical Wagner model which is based on thin airfoil theory. The theory assumes that the slope of the lift coefficient curve is a function of the angle of attack and is equal to 2π. However, for real airfoils, the slope is no longer equal to the theoretical coefficient.

A modification of each method was implemented, and a new model was suggested by modifying and combining Wagner's model (Wagner, 1924) and Goman-Khrabrov's model (Goman and Khrabrov, 1994). The modification is based on adjusting the instantaneous lift coefficient slope from 2π to the one received from static experiments in Wagner's and Goman-Khrabrov's models. As can be seen in figure 2, in the case of separated flow, this modification improved the accuracy of each method substantially, and the combination of the two models gives better results than the modified Wagner and Goman-Khrabrov models. Furthermore, we study the unsteady Kutta condition with an increased turbulence level, and when a laminar separation bubble is present. The validity of thin airfoil theory and the Kutta condition is examined by comparing the theoretical methods based on circulation and thin airfoil theory to the experimentally measured values. We determine the lift coefficient and circulation using the following methods: (i) Wagner's model (Wagner, 1924), (ii) the Kutta-Joukowski theorem (Kutta, 1902; Joukowsky, 1910) (via the circulation), (iii) Goman-Khrabrov's model (Goman and Khrabrov, 1994), and (iv) pressure measurements (Lesovoy *et al.*, 2024) at the trailing edge.

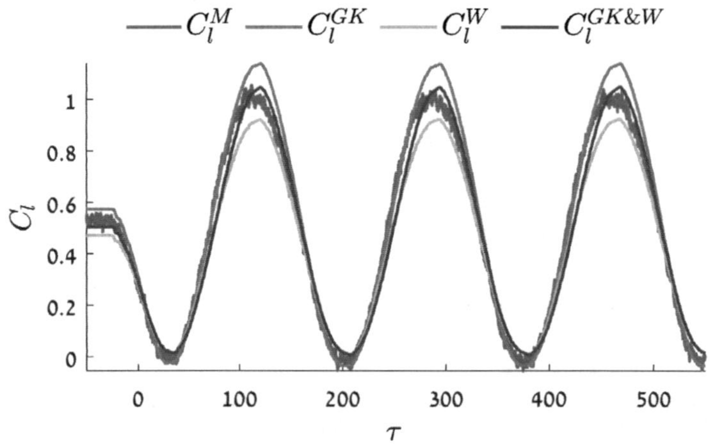

Figure 2: Comparison of lift coefficient calculated using Wagner's model, Goman-Khrabrov's model, combined model, and using pressure measurements in separated flow.

Furthermore, The examination of the Kutta condition revealed that in the case of the attached flow, the comparison between the four methods was satisfactory, indicating the validity of the Kutta condition for all models. However, in the separated flow, the unsteady Kutta condition was invalid despite the good match of the experimentally measured lift coefficient to the theoretical models.

This study provides new insights on theoretical lift coefficient models and the unsteady Kutta condition.

References

Goman M., Khrabrov A. (1994). State-space representation of aerodynamic characteristics of an aircraft at high angles of attack. *Journal of Aircraft 31*(5), 1109–1115.

Joukowsky N. (1910). Über die konturen der tragflächen der drachenflieger. *Zeitschrift für Flugtechnik und Motorluftschiffahrt 1*(22), 281–285.

Kutta W. (1902). Auftriebskräfte in strömenden flüssigkeiten. *Illustrierte Aeronautische Mitteilungen 6*(133), 133–135.

Lesovoy I., Efrati R., Stalnov O. (2024). Experimental investigation of instantaneous lift on naca 0018 airfoil section due to a non-harmonic pitching motion. *Journal of Fluids and Structures 125*, 104003.

Wagner H. (1924). Über die entstehung des dynamischen auftriebes von tragflügeln. *Z. Angew. Math. Mech. 5*, 17–35.

Assessment of Rotor Tonal Noise in the Time and Frequency Domains with Small in Diameter Propellers

Asaf Kor[*1] **and Oksana Stalnov**[2]

[1]School of Mechanical Engineering, Faculty of Engineering, Tel-Aviv University, Israel
[2]Faculty of Aerospace Engineering, Technion — Israel Institute of Technology

*contact: Asafkor47@gmail.com

The aeroacoustic signatures of propellers generated by rotating machines present a complex mix of tonal and broadband components. While rotor noise modeling has predominantly focused on the frequency domain, this study addresses the underexplored time domain model, which is crucial for precise temporal considerations and accurate near-field predictions. In contrast to the frequency domain's limitations, the time domain model enables the calculation of the full spectra of frequencies. This study reconstructs pressure signals based on instantaneous aerodynamic load measurements in the time domain, essential for handling non-periodic loads subject to random rotational speed modulations.

The acoustic signature of the rotating blade includes tonal noise in multiples of the blade passing frequency (BPF) and broadband noise. Tonal noise is attributed to steady aerodynamic loads, and to the periodic variation of loads Ffowcs Williams and Hawkings (1969); Goldstein (1976). The latter find their origin in deviation from pure axisymmetry of the flow through the rotor disk, referred to in the literature as cyclo-stationary distortions. Periodic unsteady loads arise from the non-homogeneity of the flow field encountered by the blades.

Figure 1 shows the apparatus of the study, where a rotor was placed inside an anechoic chamber and surrounded by a microphone array of $R = 1.5m$. The microphones were equally spaced along the arc with $\Delta\theta = 7.5^{\circ}$ where $\theta = 0^{\circ}$ is located above the propeller hub.

This study uses the formulation derived by Lowson in the time domain Lowson (1965), which describes the sound radiated by the unsteady singularity loads on the rotor in the time domain. This reduces the distributed loading along the blade to an equivalent point force. Radiated sound from a propeller may be considered in the circular motion of a force acting on a point at a distance r_s from the hub. The acoustic

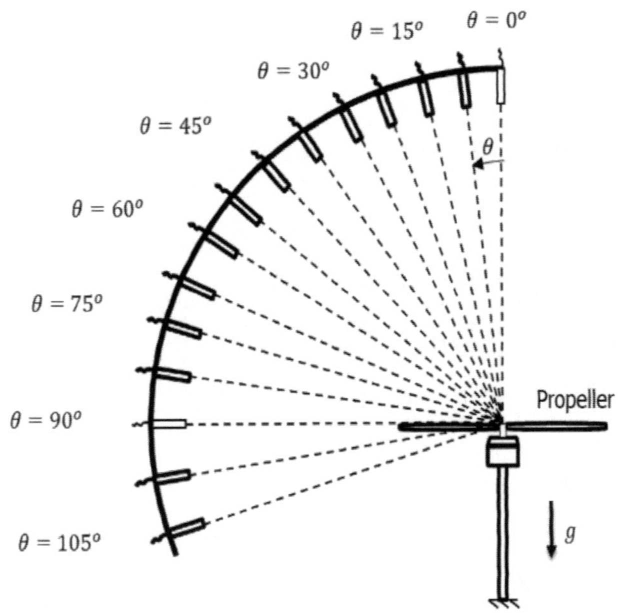

Figure 1: The experimental setup of the problem

pressure $p(t,\vec{x})$ of a rotating isolated point force in motion takes the form of

$$p(t,\vec{x}) = \frac{1}{4\pi[1-M_r]}\left[\frac{(x_i-y_i)}{c_0 r^2}\frac{\partial F_i}{\partial t}\right] + \frac{1}{4\pi[1-M_r]}\left[\frac{(x_i-y_i)(x_j-y_j)}{c_0 r^3}\frac{\partial}{\partial t}\left(\frac{M_j F_i}{1-M_r}\right)\right] \quad (1)$$

where x_i and $y_i(t)$ are the instantaneous Cartesian coordinates of the observer and the source, respectively. Here, $r(t)$ is the instantaneous absolute distance from the microphone to the location of action of the force on the blade. Square brackets indicate that the parameters are evaluated at the retarded time $\tau = t - r/c_0$. The point force is considered at the aerodynamic centre of the blade approximately at

$$r_s = \frac{2}{3}\frac{(R_T^3 - R_H^3)}{(R_T^2 - R_H^2)} \cdot R_T \quad (2)$$

with R_H and R_T as the hub and tip blade radius, respectively.

Figures 2 (a) and (b) show a PSD comparison between the measured acoustic signal in the microphone at direction $\theta = 30°$ (p^m), and the calculated signals using the presented time (p^t) and its subsequent frequency domain approximation (p^f) at different rotation rates. It can be observed that the full spectra recreation by the time domain model is accurate in its blade loading harmonics peaks values and shape predictions, and yields superior results when compared to the frequency domain results.

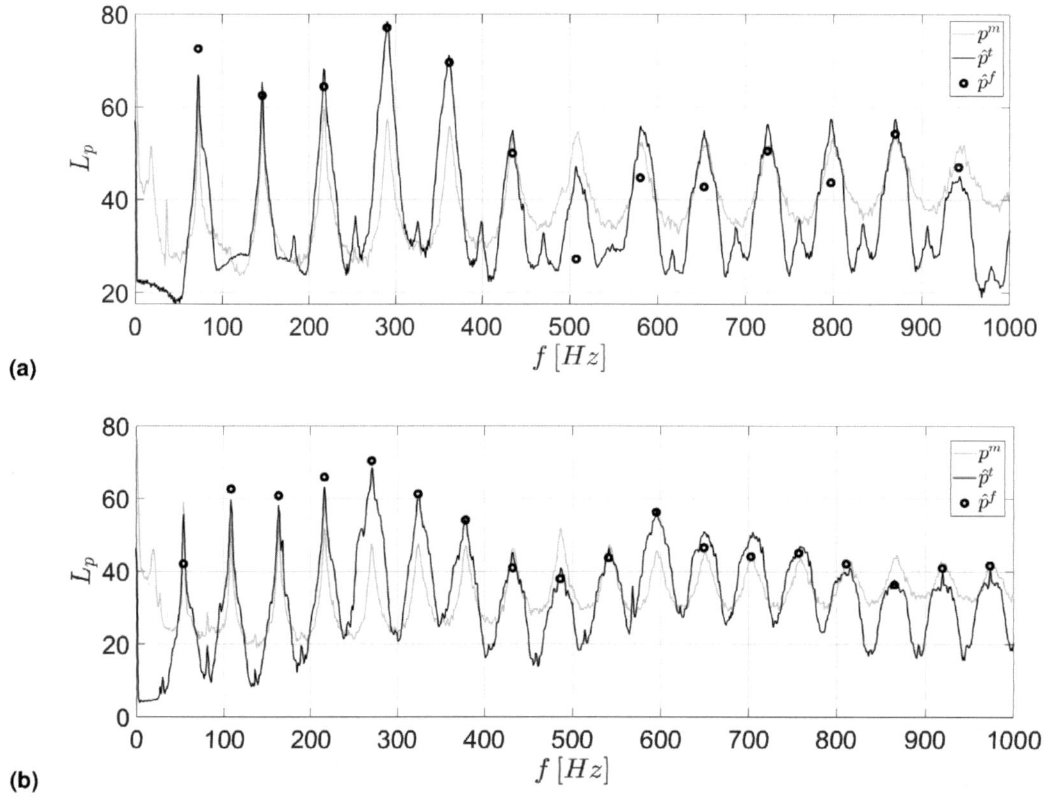

Figure 2: Comparison of the power spectral density of the measured pressure signal p^m to the time $\hat{p}^t$ and frequency $\hat{p}^f$ domain reconstructions at at direction of $\theta = 30°$ and rate of rotation (a) $N = 2175$ (b) $N = 1625RPM$.

References

Ffowcs Williams J. E., Hawkings D. L. (1969). Sound generation by turbulence and surfaces in arbitrary motion. *Philosophical Transactions of the Royal Society of London. Series A, Mathematical and Physical Sciences 264*(1151), 321–342.

Goldstein M. E. (1976). Aeroacoustics. *New York*.

Lowson M. (1965). The sound field for singularities in motion. *Proceedings of the Royal Society of London. Series A. Mathematical and Physical Sciences 286*(1407), 559–572.

PRELIMINARY STRUCTURAL DESIGN TOOL FOR FLEXIBLE SLENDER BODIES WITH AEROELASTIC CONSTRAINTS

Yaara Karniel*, Daniella E. Raveh

Technion – Israel Institute of Technology,
Technion City, Haifa, 3200003
Israel

ABSTRACT

Recent design trends in slender-body flight vehicles emphasize enhancing performance and cost reduction, resulting in increased flexibility and excessive aeroelastic phenomena. Within flexible flight vehicles, static aeroelasticity causes body deformations, altering the local angle of attack, and influencing aerodynamic loads and load distribution. The resultant flexible load distribution shifts the aerodynamic center, impacting overall aerodynamic efficiency, stability, and the control system response. Hodges [1] addressed challenges of aeroelasticity in flexible slender bodies, particularly the reduction or loss of aerodynamic stability margin. In the preliminary design phase of slender bodies, vehicle performance and load-bearing capacity take precedence, often overlooking the impact of static aeroelasticity. This stems from the conflict between the need for quick design assessment and the time-consuming aeroelastic analysis.

The study introduces the Aeroelastic Initial Design Tool (AIDTool), a computational framework for rapid assessment of the static aeroelastic response and stability of a flexible slender body. Designed for rapid performance evaluation in the preliminary design phase, AIDTool formulates aeroelastic equations of motion, calculating deflections and stability margin changes under typical flight conditions. The methodology builds on analytical structural modeling, utilizing the Assumed Modes approach [2], together with the known aerodynamic load distribution over the rigid vehicle. The latter can be computed by aerodynamic models of different fidelity in commercially available codes.

AIDTool offers easy control over parameters such as material selection, number of body segments, envelope thickness, structural openings, payload mass distribution, and more, providing a versatile approach to structural optimization. AIDTool supports parametric sensitivity study of structural parameters and an automated optimization process that relies on multiple response computations rather than gradients.

In the study, the AIDTool is validated via a static aeroelastic stability study of a slender vehicle composed of two sections of different structural properties, shown in Figure 1. The static elastic deformations computed by AIDTool (Figure 1) are in very good agreement with those computed by the ZAERO aeroelastic analysis based on a detailed finite element model [3]. The full paper will present AIDTool's structural and aerodynamic modeling approaches and a test case of preliminary structural design accounting for stability margin constraint.

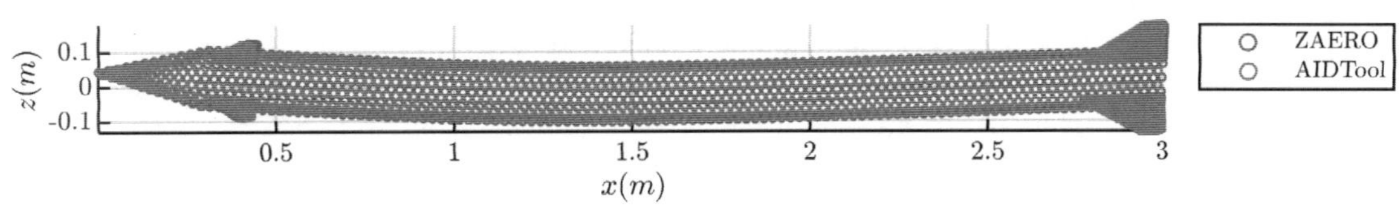

Figure 1 – Elastic Deformations of a slender body configuration as computed by AIDTool and by ZAERO

Bibliography

[1] D. H. Hodges, "A New Approach to Aeroelastic Response, Stability and Loads of Missiles and Projectiles," US Army Research Office, School of Aerospace Engineering, Georgia Institute of Technology, 2004.

[2] S. S. Rao, Mechanical Vibrations, Singapore: Prentice Hall, 2011.

[3] ZAERO Theoretical Manual, ZONA Technology Inc., 2019.

Flutter and Post-Flutter Response of Plates with Acoustic Cavity Coupling

Maxim Freydin *
Technion – IIT, Haifa, Israel, 3200003

Abstract

A traditional flutter analysis consists of an elastic structure and a freestream flow where the key question is at which freestream flow condition does flutter occur. Beyond flutter onset the coupled fluid-structure system responds with a limit cycle oscillation which can be characterized by a spatial shape and frequency content. This is of course true for elastic structures to which flutter onset is not destructive. Thin elastic plates are a typical example of such structures. Another important type of analysis is acoustic noise transmission from external pressure fluctuation to a closed volume (e.g., cavity) through an elastic wall. The geometry, boundary conditions, and material properties of the elastic section of a cavity, as well as the geometry and fluid properties inside the cavity, all affect the transmission of external noise. The two analyses are almost always exclusive in the literature [1]. For example, when the flutter onset condition is of interest acoustic effects are neglected or simplified significantly by adjusting the effective mass and stiffness properties of the elastic structure. This work presents early results that demonstrate the importance of dynamic coupling between the plate, acoustic cavity, and freestream flow for predicting flutter onset and the properties of the post-flutter response. Computational and theoretical results are presented with a correlation to a recent experiment conducted at North Carolina State University (NC State). It is shown that the dynamic coupling between all elements of the analysis is necessary for reaching flutter onset within the stagnation pressure available at the NC State supersonic wind tunnel.

The aerodynamic and structural response of a thin elastic plate at Mach 2.5 in post-flutter oscillation was measured in recent wind tunnel experiments conducted at NC State [2]. A computational study is currently in progress investigating the effect of dynamic coupling between the plate and cavity on the linear stability and nonlinear post-flutter response of the fluid-structure system. In this work, results show that a fully coupled analysis is necessary for accurate prediction of the flutter onset condition and the properties of the post-flutter limit cycle oscillation. A schematic of the test article is shown in Figure 1 consisting of a thin elastic plate with supersonic freestream flow on one side, a closed pressurized cavity on the other, and two piezoelectric elements. Full-field measurements of static pressure were acquired at 10 kHz using pressure-sensitive paint and synchronized with the measured voltage of the piezoelectric elements to allow simultaneous analysis of the aerodynamic and structural response. A computational model is presented to analyze the fluid-structure system including the effect of rivet connection as locally elastic in-plane boundary conditions. An early theoretical analysis of a reduced 3 degrees-of-freedom system (one cavity and two structural modes) is shown to capture the effect of cavity-facilitated flutter onset of a plate.

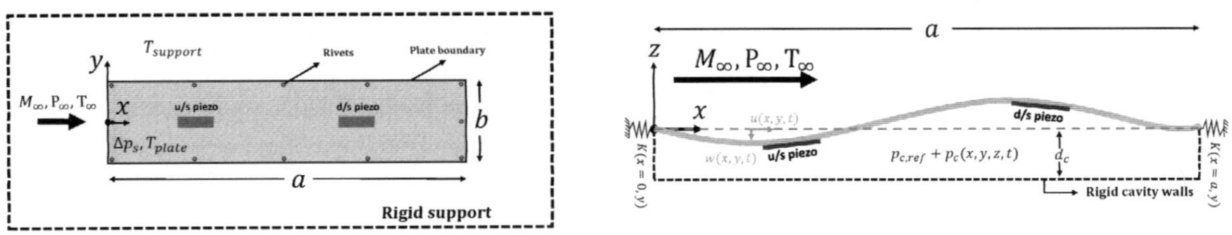

(a) **(b)**

Fig. 1 (a) Top and (b) side views of an elastic plate with two piezoelectric patches, freestream flow, cavity, static pressure differential, in-plane edge stiffness, rivets, and support and plate temperatures.

*Lecturer, Aerospace Engineering, freydinmaxim@technion.ac.il

References

[1] Dowell, E. H., *Aeroelasticity of plates and shells*, Noordhoff International Publishing (now Springer), 1974.

[2] Varigonda, S. V., Serafim, L. P., Freydin, M., Dowell, E. H., and Narayanaswamy, V., "Two-dimensional pressure field imaging of an elastic panel executing post-flutter oscillations," *Journal of Fluids and Structures*, Vol. 125, 2024, p. 104056.

Hypersonic Vehicle Aeroelasticity Research Using Ansys Aerodamping Workflow

Dvir Mendler* and Konstantinos Giannokostas*
Ansys, Israel and Ansys, Greece

I. Abstract

Vehicles traveling within a hypersonic flow regime may encounter elastic instabilities due to extreme flow conditions. In this study the intake of a scramjet engine has been addressed since its tight connection between flow and structure can be crucial to the operation of the vehicle.

Research and development of hypersonic vehicles can be very expensive and difficult to conduct using experiments due to the complexity of creating hypersonic flow environment. Computational methods are therefore extremely relevant. Aeroelasticity is usually modeled using two-way fluid-structure interaction (two-way FSI) which can effectively simulate flutter in computational simulation.

Two-way FSI is a computational expensive method which requires large computation resources and may take a long time to run. To save runtime and make flutter prediction more accessible for engineering tasks, Ansys developed the "Aerodamping workflow". In the "Aerodamping workflow" a modal analysis done in "Ansys Mechanical" combined with flow analysis in "Ansys Fluent" can predict rather a structure will flutter in a specific flow regime. That one-way coupling can be considerably faster than traditional highly coupled two-way FSI. In other words, the simulation results will show rather the geometry is aerodynamically stable (damped) or not, using a computationally effective simulation.

In this work a comparison was made to a work done by (Bhattrai, Neely, Currao, & McQuellin, 2019). In their paper the authors investigated a hypersonic vehicle intake geometry using experimental and computational tools. A specific configuration where geometry deflections are extreme was chosen from the paper, and investigated in this work with two-way FSI simulation in first stage and using the "Aerodamping workflow" in second to compare the results and computation costs between the two different methods and both works.

Results of this work show that it is possible to know rather a body will be aerodynamically stable with a computational affordable, loosely coupled simulation, using the Ansys "Aerodamping workflow".

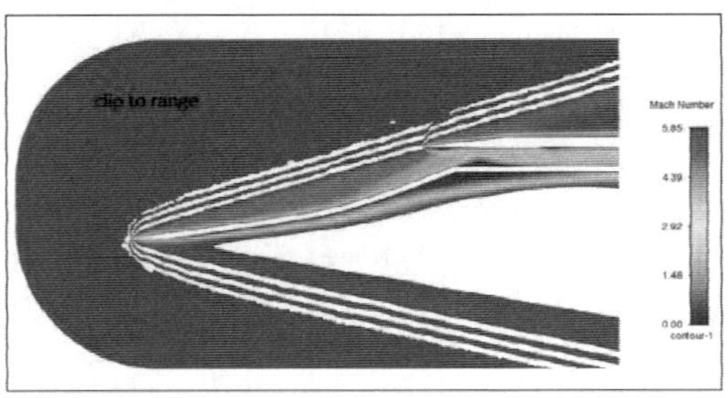

* Senior application engineer, Fluids, dvir.mendler@ansys.com

Flight-Dynamics Aeroelastic Coupling in Flexible and Very Flexible Wings

Dor Naftaly* and Daniella E. Raveh†

I Introduction

Aircraft designs are becoming more flexible than ever before. This is the case for transport, fighter jets and unmanned aerial vehicles (UAVs). While the common practice in production aeroelastic analyses is to neglect the inertial coupling between the elastic and rigid body motion, when the aircraft undergoes large deformations or hard maneuvers, the coupling terms might have a significant effect on the aircraft's loads and stability envelope (see for example the Helios mishap [1]) and might not be negligible.

In recent years, several formulations of the coupled flight-dynamics aeroelastic EOM were offered in the literature [2–4]. They vary in the selection of the coordinate system, assumptions made in the derivation, and the choice of nonlinear modeling of the wing's structure in large deformations.

II Work Description

The current study presents an analytical derivation of the coupled flight dynamics aeroelastic equation of motion (EOM), in a body coordinate system, for linearly elastic and very flexible aircraft that undergo hard maneuvers. The study provides an analytical assessment of coupling terms' magnitude and their significance. The inertial coupling effects of a general flexible flying wing are investigated.

III Preliminary Results

Mathematical models were derived for the coupled nonlinear flight-dynamics aeroelastic system, which are valid for both small and large elastic deformations. The EOM system includes an equation set for the transnational motion of the center of gravity (CG), an equation set for the rotational motion of the CG, and an equation set for the elastic motion of the structural DOFs with respect to the CG. These three equation sets can be stated in a matrix form as follows: [a]

$$\begin{bmatrix} m\underline{I} & \underline{0} & \underline{\widetilde{I}}^T \cdot \underline{M} \\ \underline{0} & \underline{I}_M & -\underline{R}_b^T \cdot \underline{M} \\ \underline{M} \cdot \underline{\widetilde{I}} & -\underline{M} \cdot \underline{R}_b & \underline{M} \end{bmatrix} \cdot \begin{bmatrix} \dot{\vec{V}}_0 \\ \dot{\vec{\omega}} \\ \ddot{\vec{u}}_e \end{bmatrix} = \begin{bmatrix} R\vec{H}S_L \\ R\vec{H}S_A \\ R\vec{H}S_E \end{bmatrix} \tag{1}$$

*Ph.D. Student, Faculty of Aerospace Engineering, dor.naf01@gmail.com
†Professor, Faculty of Aerospace Engineering, daniella@technion.ac.il
[a] The terms will be outlined in the full paper, due to the word count limit

where:

$$
\begin{aligned}
R\vec{H}S_L &= -\vec{\omega} \times \left(m\vec{V}_0 + \underline{\widetilde{I}}^T \cdot \underline{M} \cdot \dot{\vec{u}}_e \right) + \underline{\widetilde{I}}^T \cdot \vec{F}_{ex} + m\vec{g} \\
R\vec{H}S_A &= -\vec{\omega} \times \left(\underline{I}_M \cdot \vec{\omega} - \underline{R}_b^T \cdot \underline{M} \cdot \dot{\vec{u}}_e \right) \\
&\quad - \dot{\underline{I}}_M \cdot \vec{\omega} + \dot{\underline{R}}_b^T \cdot \underline{M} \cdot \dot{\vec{u}}_e - \underline{R}_b^T \cdot \vec{F}_{ex} \\
R\vec{H}S_E &= -\left(\underline{M} \cdot \underline{\widetilde{\Omega}} + \underline{\widetilde{\Omega}} \cdot \underline{M} \right) \cdot \dot{\vec{u}}_e - \underline{\widetilde{\Omega}} \cdot \underline{M} \cdot \underline{\widetilde{I}} \cdot \vec{V}_0 - \underline{\widetilde{\Omega}} \cdot \underline{M} \cdot \underline{\widetilde{\Omega}} \cdot \vec{r}_b \\
&\quad + \vec{F}_{ex} + \underline{M} \cdot \underline{\widetilde{I}} \cdot \vec{g} - \underline{K} \cdot \vec{u}_e
\end{aligned}
\tag{2}
$$

It is evident that while the transnational and rotational equation sets are lightly coupled (only via non-acceleration terms), both are strongly coupled with the elastic equation set. The coupled EOM provides for time-domain simulations of initial conditions response, and stability analyses.

Table 1 lists the substantial coupling effects and provides their magnitudes as a function of the out-of-plane elastic deformations, u_z, the frequency of the aeroelastic first symmetric bending mode, ω_n, the aircraft wing span, b, and the maneuver rates and accelerations.

Term	Impact	Magnitude
$\vec{\omega} \times \left(\underline{R}_b^T \cdot \underline{M} \cdot \dot{\vec{u}}_e \right)$	pitching moment	$mbr\omega_n u_z$
	yawing moment	$mbq\omega_n u_z$
$\dot{\underline{I}}_M \cdot \vec{\omega}$	pitching moment	$mbr\omega_n u_z$
	yawing moment	$mbq\omega_n u_z$
$\underline{M} \cdot \underline{\widetilde{I}} \cdot \dot{\vec{V}}_0$	elastic loads	mgn_z
$\underline{\widetilde{\Omega}} \cdot \underline{M} \cdot \underline{\widetilde{I}} \cdot \vec{V}_0$	elastic loads	mqV
$\underline{M} \cdot \underline{R}_b \cdot \dot{\vec{\omega}}$	elastic loads	$mb\dot{p}$
$\underline{\widetilde{\Omega}} \cdot \underline{M} \cdot \underline{\widetilde{\Omega}} \cdot \vec{r}_{b0}$	elastic loads	$mbpq$
$\underline{K} \cdot \vec{u}_e + \underline{\widetilde{\Omega}} \cdot \underline{M} \cdot \underline{\widetilde{\Omega}} \cdot \vec{u}_e$	elastic loads	$m\omega_n^2 \left(1 - \frac{q^2+p^2}{\omega_n^2}\right) u_z$

Table 1. Flight-dynamics-aeroelastic coupling effects and their magnitudes

The full paper will include a discussion of the significant coupling terms for cases of linear elastic / nonlinear large deformations and in light/hard maneuvers, and will investigate the inertial coupling effects with a general flexible flying wing configuration test case.

References

[1] Noll, T. E., Brown, J. M., Perez-Davis, M. E., Ishmael, S. D., Tiffany, G. C., and Gaier, M., "Investigation of the Helios prototype aircraft mishap volume I mishap report," Vol. 9.

[2] Murua, J., Palacios, R., and Graham, J. M. R., "Applications of the unsteady vortex-lattice method in aircraft aeroelasticity and flight dynamics," *Progress in Aerospace Sciences*.

[3] Meirovitch, L., "Hybrid state equations of motion for flexible bodies in terms of quasi-coordinates," *Journal of Guidance, Control, and Dynamics*.

[4] Su, W. and Cesnik, C. E., "Nonlinear aeroelasticity of a very flexible blended-wing-body aircraft," *Journal of Aircraft*.

Wind-Tunnel Investigation of the F-16 Transonic Buffet Phenomenon

Maj. Tzlil Nahom Jidovetski* and Maj. Michael Iovnovich† and Daniella E. Raveh‡

I Introduction

Buffet refers to a group of unsteady aeromechanic phenomena in which flow instabilities around the aircraft lifting surfaces lead to oscillatory forces which may degrade the platform handling qualities and structural integrity characteristics. While buffet originates from a flow phenomenon, its characteristic frequencies often fall within the range of an aircraft's structural dynamics natural frequencies, which may lead to dynamic loads magnification by means of aeroelastic resonance interaction. Consequently, buffet is a significant design factor and certification challenge in flight envelope expansion, structure strength substantiation, and fatigue life assessment processes for both commercial and fighter platforms.

The current study is focused on transonic buffet, also termed shock-buffet, which is a type of a buffet phenomenon that is characterized by self-sustained shock-wave oscillations which occur for a range of mean-flow Mach numbers and moderate angles of attack due to an interaction between the shock and a partially or fully separated boundary layer. The phenomenon has been extensively investigated in the literature, both numerically [1–7] and in wind tunnel experiments [8–11] since the 1980s.

Previous shock-buffet studies on fighter jets such as the F-104 [12] and F-111 [13] are very limited in the public domain and primarily involve instrumented flight test measurements. The F-35 aircraft is an example of a fighter platform that exhibits multiple buffet phenomena including transonic wing buffet, which had a significant impact on its structural design and certification processes [14]. Predictive characterization and quantification of buffet loads are essential for the structural design purposes of such configurations. In the absence of an alternative, reliable buffet load models may become available only at the late stages of the design process, based on costly flight test surveys using an instrumented aircraft. Therefor, there is a strong motivation to gain more knowledge of the transonic buffet phenomenon on fighter aircraft, as well as to develop reliable and efficient buffet loads assessment strategies using CFD and wind tunnel tests.

II Work Description

This study is conducted as part of multi-year effort by the IAF to develop and validate a buffet-load prediction capability primarily for the purpose of enhancing fighter aircraft external stores certification processes. The current study is focused on a wind-tunnel investigation of the F-16 transonic buffet phenomenon. The test was conducted on a rigid (metalic) F-16 model (1:15) instrumented with 14 unsteady pressure transducers

*Head of Flight Sciences Group, tzlilnahom@gmail.com
†Lead Aeroelasticity Engineer, Flight Sciences Group, michaeliovnovich@idf.il
‡Professor, Faculty of Aerospace Engineering, daniella@technion.ac.il

at transonic flow conditions. Sensor locations are schematically presented in figure 1. These locations were selected based on the expected shock travel region available from numerical predictions. The first goal of this test was to identify and map the F-16 buffet envelope in terms of Mach number and angles of attack. Once developed buffet conditions were identified, unsteady pressure data was collected at various buffet conditions to characterize the phenomenon and compare its physical mechanism and characteristics with those previously known from transport aircraft buffet studies. A comparative study was then conducted for various external store downloadings in order to assess the effects of external stores at different store stations on wing buffet characteristics.

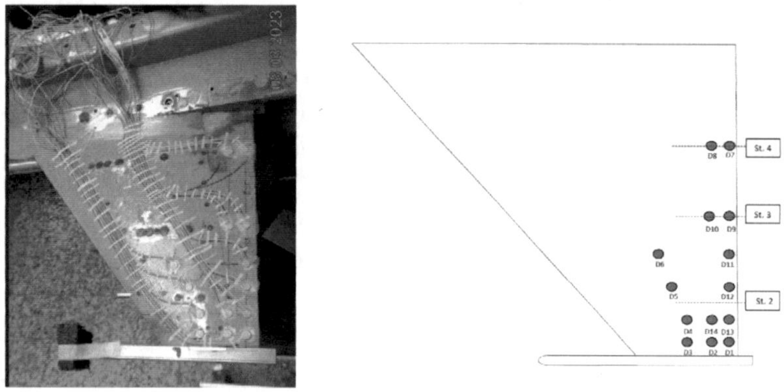

Figure 1. Dynamic pressure transducers distribution over the wing

III Preliminary Results

Preliminary results are presented for both clean store stations and loaded store stations (with missiles/stores) configurations. Figure 1 shows a comparison between two unsteady surface-pressure RMS maps over the upper wing surface for both configurations. Pressure sensor locations are denoted by red points. RMS levels were calculated based on pressure time history data obtained at flow conditions of $\alpha = 12.7°$ and Mach 0.85. High unsteady pressure RMS values appear in areas of high shock motion and flow separation/reattachment. In this comparison, the presence of underwing stores tends to decrease buffet intensity and thereby to reduce the resultant unsteady wing loads.

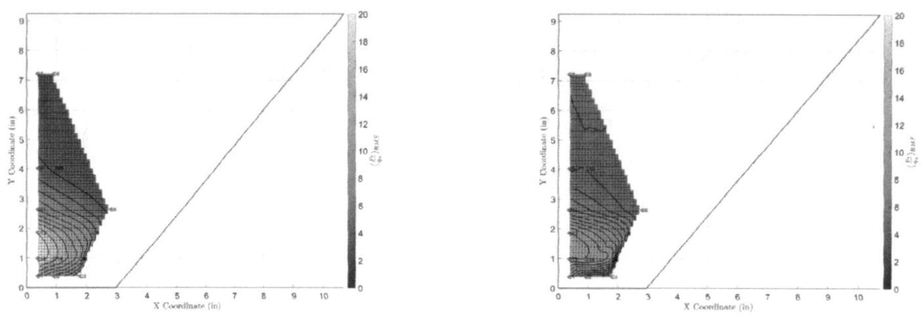

(a) Clean store stations configuration (b) Loaded store stations Configuration

Figure 2. Surface-pressure rms levels at $\alpha = 12.7°$ and Mach 0.85, developed buffet conditions

References

[1] Iovnovich, M. and Raveh, D. E., "Numerical Study of Shock Buffet on Three-Dimensional Wings," *AIAA Journal*, Vol. 53, No. 2, 2015, pp. 449–463.

[2] Iovnovich, M. and Raveh, D. E., "Reynolds-Averaged Navier-Stokes Study of the Shock-Buffet Instability Mechanism," *AIAA journal*, Vol. 50, No. 4, 2012, pp. 880–890.

[3] Lee, B., "Self-Sustained Shock Oscillations on Airfoils at Transonic Speeds," *Progress in Aerospace Sciences*, Vol. 37, No. 2, 2001, pp. 147–196.

[4] Poplingher, L. and Raveh, D. E., "Comparative Modal Study of the Two-Dimensional and Three-Dimensional Transonic Shock Buffet," *AIAA Journal*, Vol. 61, No. 1, 2023, pp. 125–144.

[5] Plante, F., Dandois, J., and Laurendeau, É., "ZDES and URANS Simulations of 3D Transonic Buffet Over Infinite Swept Wings," in "Progress in Hybrid RANS-LES Modelling," Springer, pp. 259–269, 2020.

[6] Sartor, F. and Timme, S., "Reynolds-averaged Navier-Stokes simulations of shock buffet on half wing-body configuration," in "53rd AIAA Aerospace Sciences Meeting," , 2015, p. 1939.

[7] Kidron, Y., Levy, Y., and Wasserman, M., "The EZAir CFD Solvers Suite: EZAir Theoretical and User's Manual-Version 3.54," Tech. rep., Tech. Rep. August, IS-CFDC, 2016.

[8] Jacquin, L., Molton, P., Deck, S., Maury, B., and Soulevant, D., "Experimental Study of Shock Oscillation Over a Transonic Supercritical Profile," *AIAA journal*, Vol. 47, No. 9, 2009, pp. 1985–1994.

[9] Dandois, J., "Experimental Study of Transonic Buffet Phenomenon on a 3D Swept Wing," *Physics of Fluids*, Vol. 28, No. 1, 2016, p. 016101.

[10] Koike, S., Ueno, M., Nakakita, K., and Hashimoto, A., "Unsteady pressure measurement of transonic buffet on NASA common research model," in "34th AIAA Applied Aerodynamics Conference," , 2016, p. 4044.

[11] Lee, B., Ellis, F., and Bureau, J., "Investigation of the Buffet Characteristics of Two Supercritical Airfoils," *Journal of aircraft*, Vol. 26, No. 8, 1989, pp. 731–736.

[12] Friend, E. L. and Sefic, W. J., "Flight Measurements of Buffet Characteristics of the F-104 Airplane for Selected Wing-Flap Deflections," Tech. rep., NASA TN D-6943, 1972.

[13] Monaghan, R. C., "Flight-Measured Buffet Characteristics of a Supercritical Wing and a Conventional Wing on a Variable-Sweep Airplane," Tech. rep., NASA-TP-1244, 1978.

[14] Ellis, R. M., Gross, P., Yates, J. B., Casement, J. R., Chichester, R. H., and Nesmith, K., "F-35 Structural Design, Development, and Verification," in "2018 Aviation Technology, Integration, and Operations Conference," , 2018, p. 3515.

IACAS

Flutter Control of a Symmetric Wing Using Active Flow Control

Shay Monat*[1] and Oksana Stalnov[2]

[1,2,3]School of Mechanical Engineering, Faculty of Engineering, Tel-Aviv University, Israel
[4]Faculty of Aerospace Engineering, Technion — Israel Institute of Technology

*contact: Shaymonat@gmail.com

A wind tunnel experimental investigation of self-sustained oscillations of aeroelastic symmetric Glauert airfoil section is studied. The wing is free to rotate around its angle of attack axis (see Figure 1). An active flow control (AFC) system in the form of blowing jets is used to control the oscillations. Several open and closed loop control laws are tested to evaluate their effect on the natural fluttering of the wing.

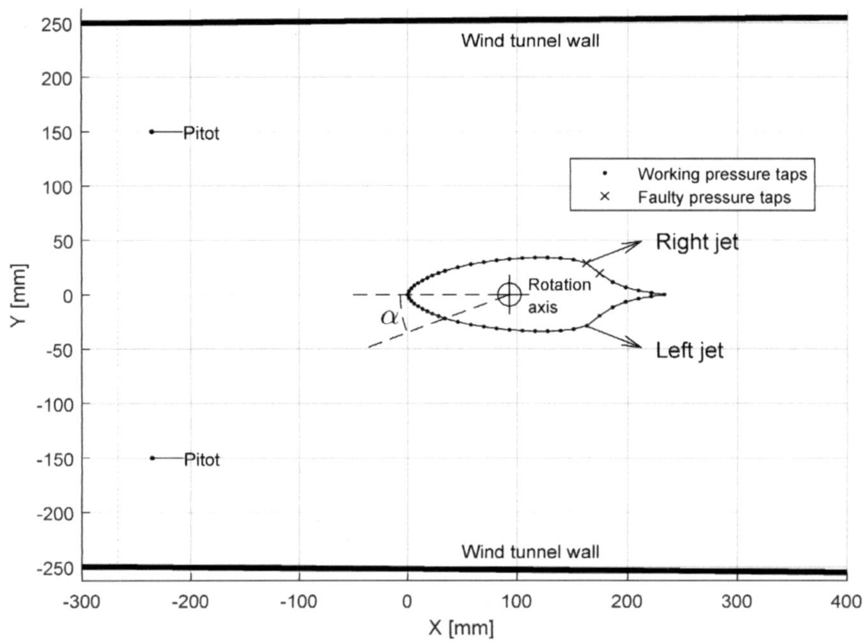

Figure 1: Schematic description of the symmetric Glauert airfoil with pressure sensors coordinates.

Experimental evidence of limit-cycle-oscillation (LCO) related to the transitional flow regime exists in the wind tunnel experiments of Poirel et al. Poirel *et al.* (2008); Poirel and Mendes (2014). They revealed small-amplitude self-sustained oscillations of a NACA0012 airfoil elastically mounted in one-degree-of-freedom (1-DOF) pitch. This phenomenon was attributed to non-linearity in the aerodynamic loads provided by laminar boundary layer separation, leading to the laminar separation flutter.

Figure 2 shows measurements of the lift coefficient (C_l) and angle of attack (α) of the wing model. These results show the wing model shows the natural LCO of the wing as it is released at $t = 0[sec]$ from $\alpha = 0°$ and begins to flutter. The lift coefficient is evaluated using integration of the pressure taps around the section of the wing model (indicated in Figure 1).

The use of AFC actuators has been studied in the past in many applications (Cattafesta III and Sheplak, 2011) but very few attempts have been made to control their actuation using a closed loop control law (Troshin and Seifert, 2013). In the current setup, the blowing intensity is kept constant while the side from which the jets blow (see Figure 1) can be changed rapidly.

Several control laws have been investigated where the blowing sides have been switched by some control logic such as: constant switching frequency (open loop), angle of attack dependent algorithm (closed loop), pressure-tap dependent algorithm (closed loop) and several others. One example is shown in Figure 3, where the control law is dependent on the real-time reading of the angle of attack. In this test the control algorithm changed the blowing sides at a certain phase-lag of the angle of attack.

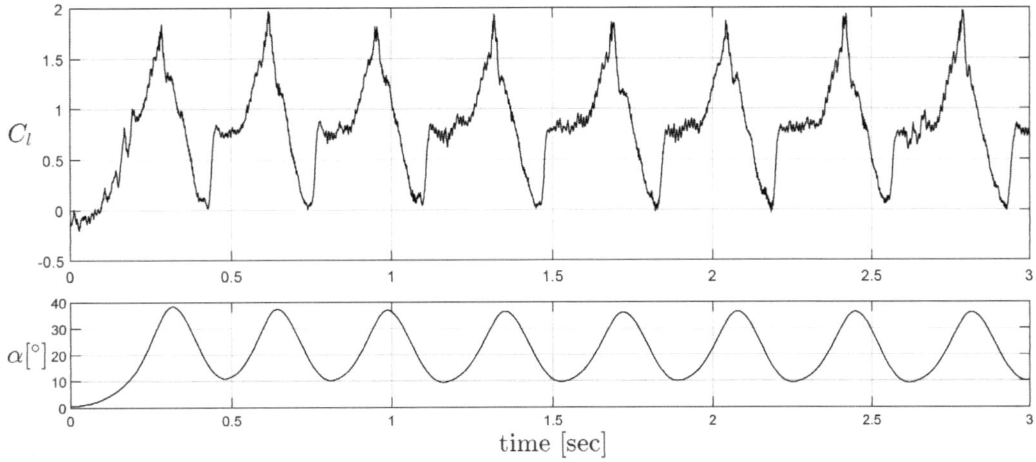

Figure 2: The lift coefficient (top) and angle of attack (bottom) with respect to time at $Re = 3.2 \times 10^5$.

This phase lag have been scanned systematically and the results show that this variable has an effect on the mean angle of attack, its amplitude and frequency.

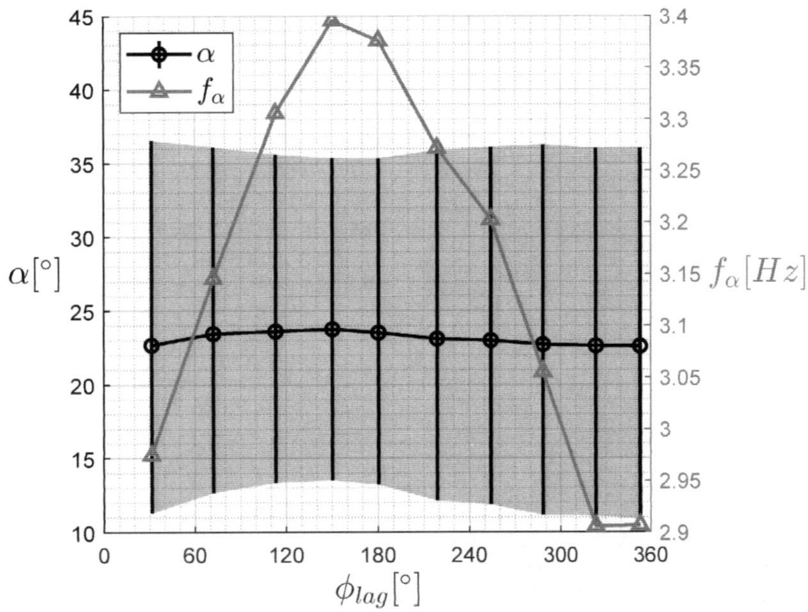

Figure 3: The mean angle of attack and its standard deviation (left ordinate) and the frequency of the angle of attack (right ordinate) with respect to the phase lag between the angle of attack and the side switching of the AFC jets. $Re = 3.2 \times 10^5$.

The full study shows a thorough investigation of various control laws and their effect on the dynamic nature of the flow around the wing. The paper will dive deeper into unsteady phenomena and discuss the insights which this setup uncovers and the possible uses of such a system in real-world applications.

References

Cattafesta III L. N., Sheplak M. (2011). Actuators for active flow control. *Annual Review of Fluid Mechanics 43*, 247–272.

Poirel D., Harris Y., Benaissa A. (2008). Self-sustained aeroelastic oscillations of a naca0012 airfoil at low-to-moderate reynolds numbers. *Journal of Fluids and Structures 24*(5), 700–719.

Poirel D., Mendes F. (2014). Experimental small-amplitude self-sustained pitch–heave oscillations at transitional reynolds numbers. *AIAA journal 52*(8), 1581–1590.

Troshin V., Seifert A. (2013). Performance recovery of a thick turbulent airfoil using a distributed closed-loop flow control system. *Experiments in fluids 54*, 1–19.

Design, numerical and experimental investigation of wing with a morphing aileron section

Gali Alon Tzezana[1], Gal Doron[1], Valentin Nov[1], Avishay Kidron[1]

***Abstract*— Ailerons are popular control surfaces used to create rolling moment. However, traditional ailerons inherently create a discontinuity in the wing surface, which can lead to turbulence, noise, and increased drag. Morphing ailerons are widely considered to reduce drag due to the smooth variation in the wing shape [1][2].**

In this work, a novel design for a morphing aileron is proposed. The morphing aileron section was fabricated and incorporated in a wing with aspect ratio 5.2, and its aerodynamic properties were evaluated numerically and tested in the transonic wind tunnel at Israel Aerospace Industries (IAI). Excellent agreement was obtained between the numerical solution and the experimental results.

I. Wing Design

The actuation concept for the morphing aileron is based on a flexible trailing edge made of composite materials. The aileron deflection is achieved by the rotation of four curved actuation rods. The rods are controlled individually by four servo motors, thus allowing a smooth variation in the deflection angle along the span and, potentially, variation of the maximum deflection location. For the wind tunnel testing, the morphing aileron was deployed on a rigid aluminum wing shown in Figure 1a.

II. Methods

The wing performance was evaluated numerically for different aileron deflections. First, the two-dimensional aerodynamic coefficients for airfoils with different flap angles were obtained using the CFD solver EZAir, developed by the Israeli CFD Center (ISCFDC) [3]. Then, performance prediction for the three-dimensional wing was calculated using a modified Weissinger's L-method [4].

In addition to the numerical prediction, wind tunnel tests were conducted in the transonic wind tunnel in IAI. During each test, deflection angles were changed in a pitch-pause manner at intervals of 1.5°. The experiments were performed at the following conditions:

- Wing sweep: $\Lambda=20°$
- Mach number: $M=0.5$
- Angle of attack: $\alpha=-2°$
- Maximum aileron deflections: $\delta_a \epsilon [-3,6]°$ (at the center of the aileron)

III. Results

The results show high repeatability, and no hysteresis was found. The normal force (Figure 2) and the rolling moment (Figure 3) due to the morphing aileron's deflection are well predicted by Weissinger's method.

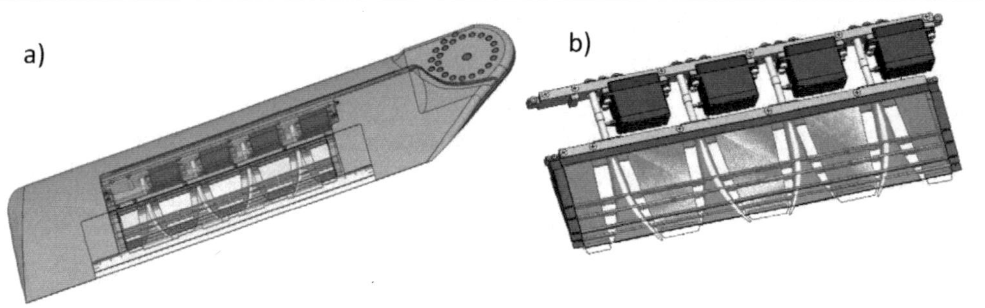

Figure 1. Schematic of the wing with a morphing aileron (a), and a detailed view of the flexible trailing edge and morphing mechanism (b).

[1] Rafael Advanced Defense Systems Ltd., POB 2250, Haifa, 3102102 Israel

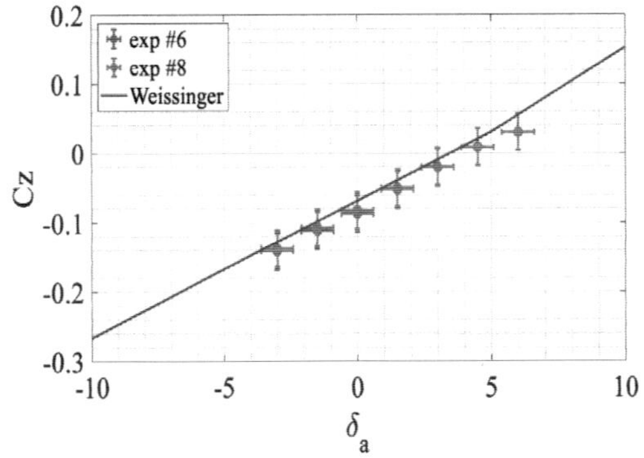

Figure 2. Normal force coefficient vs. aileron deflection - experiment and prediction

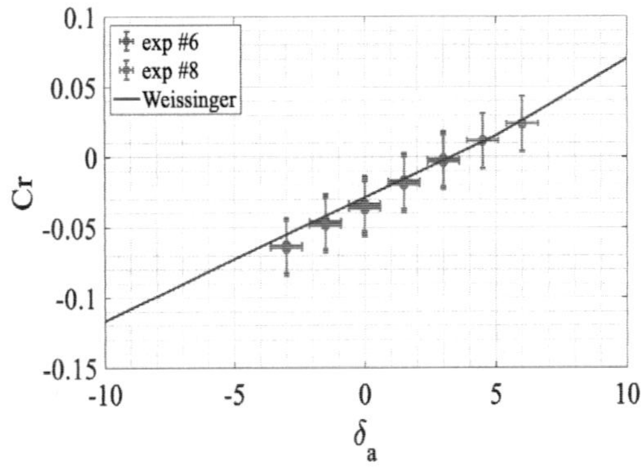

Figure 3. Rolling moment coefficeint - experiment and prediction

REFERENCES

[1] G. Amendola, I. Dimino, M. Magnifico and R. Pecora (2016). Distributed actuation concepts for a morphing aileron device. The Aeronautical Journal, 120, pp 1365-1385 doi:10.1017/ aer.2016.64

[2] De Gaspari, A., Cavalieri, V., & Fonzi, N. (2023). An energy-based design approach in the aero-structural optimization of a morphing aileron. MATERIALS RESEARCH PROCEEDINGS, 37, 337-340.

[3] Kidron, Y., Levy, Y., & Wasserman, M. (2016). The EZAir CFD Solvers Suite: EZAir Theoretical and User's Manual-Version 3.54. Tech. Rep. August, ISCFDC.

[4] Weissinger, J. (1947). The lift distribution of swept-back wings (No. NACA-TM-1120).

“The Dream Chaser” Re-Entry Space Vehicle Actuation System

Ilan Berlowitz
IBAero Engineering
iberlowitz24@gmail.com

Abstract

Since Yuri Gagarin's historic mission, re-entry missions have placed significant emphasis on ensuring the safe return of astronauts and minimizing mission costs. The re-entry phase is pivotal for these objectives, occurring within a harsh environmental context that demands meticulous vehicle design and environmental analysis.

This study aims to develop an actuation system for the vertical fin (rudder) of a re-entry space vehicle, specifically NASA's “The Dream Chaser”. It delves into the re-entry environment, trajectories, and missions while exploring various actuation systems.

The study involves calculating loads on the vertical fin and estimating deflection angles. Kinematic analysis is conducted to ascertain actuator characteristics, with a focus on evaluating power requirements. Additionally, key components are sized, and total actuator mass is estimated.

The primary considerations for a re-entry mission actuation system are mass and power specifications. While the Electro-Hydrostatic Actuator (EHA) requires relatively more power and mass compared to EMA, it was chosen due to its proven functionality under harsh environmental conditions. However, EMA technology holds potential to rival EHA, provided issues such as jamming are effectively resolved.

“The Dream Chaser”, is a reusable spaceplane designed for crewed missions to low Earth orbit (LEO). It can carry up to seven people and cargo, launching vertically on the “Vulcan Centaur” rocket and autonomously landing horizontally on runways. A cargo variant is already operational for resupplying the International Space Station (ISS).

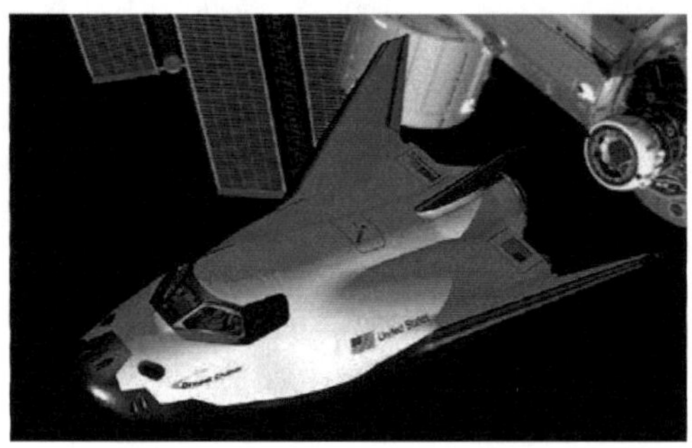

“The Dream Chaser”

"The Dream Chaser" Specifications

Crew	Up to 7
Length	9 m
Max wingspan	7 m
Height	2.8 m
Reference body length (no fins)	8.2 m
Reference body span (no fins)	4.5 m
Body performance area (no fins)	36.98 m^2
Vertical fin (rudder) area	1.877 m^2
Habitable volume	16 m^3
Mass	9,000 kg
Propellants	Liquid (C_3H_8/N_2O)
Electrical system	Batteries
Escape system	Liquid boosters
Hypersonic L/D ratio	1.8
Cross-range	2038.5 km
Landing speed	307 km/h
Glide time	~60 s

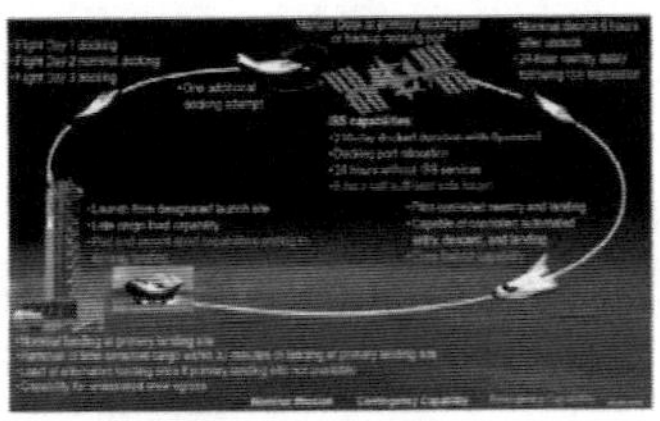

Typical Mission

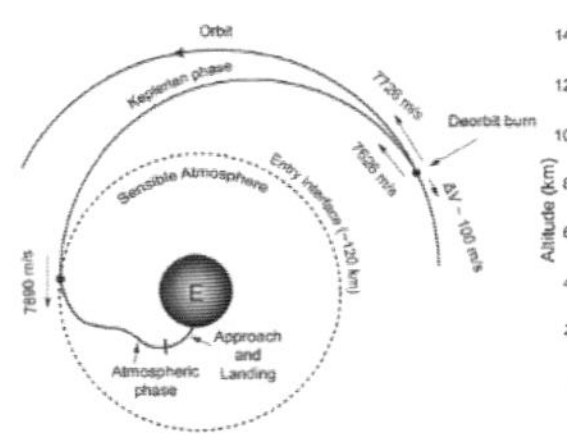

Phases of the Re-Entry

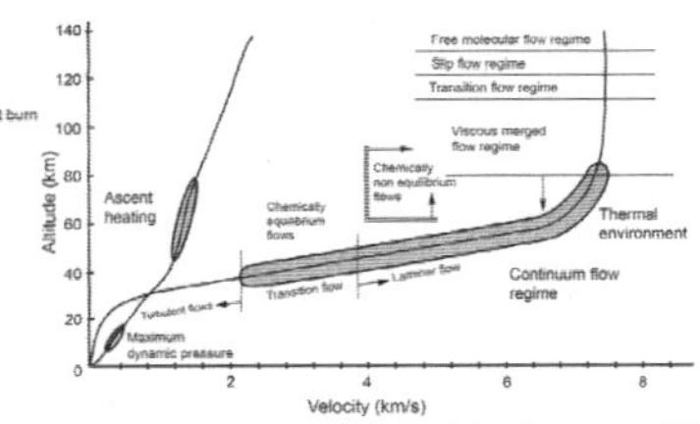

Re-Entry Flow Regimes

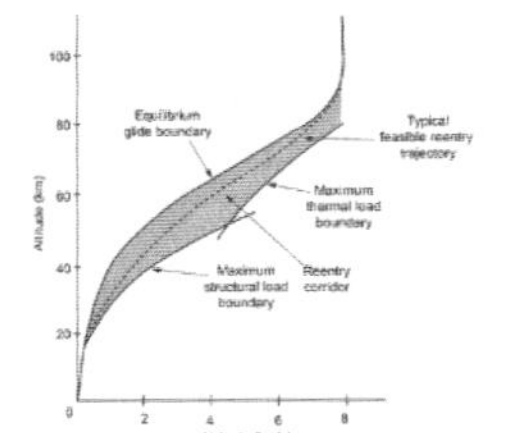

The "corridor" of the Re-Entry

Different flow regimes, depicted in the above figures, include:

- **Above 130 km:** Free Molecular Flow Regime.
- **Between 130 km and 120 km:** Slip Flow Regime.
- **Between 120 km and 110 km:** Transition Flow Regime.
- **Between 110 km and 75 km:** Viscous Merged Flow Regime.
- **Below 75 km:** Continuous Flow Regime, characterized by the highest aerodynamic loads, allowing for application of classic fluid mechanic theory, considering flat plate under hypersonic flows.

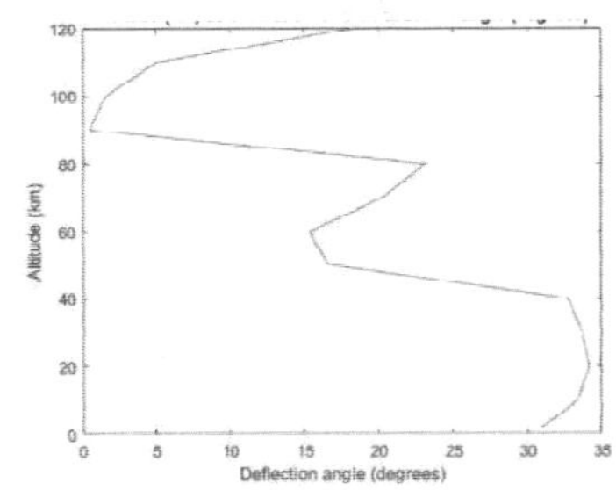

Altitude as a function of the deflection angles

The Knudsen Number allows characterizing the boundary conditions of a fluid flow. The Knudsen Number below 120 km < 0.01 → Continuous Flow Regime.

The trajectory can be divided into three distinct segments:

- A phase characterized by a steady decrease in altitude from 120 km to 90 km as the vehicle passes through the re-entry interface at a constant speed, approximately 8 km/s (~Mach 25).

- Descending from 80 km to 40 km altitude, the speed steadily decreases until reaching 0.5 km/s, with the maximum drag, forces, hinge moments and the required power values attained at 40 km altitude. **Note**: angle of attack 45° from 120 km to 40 km.
- Below 40 km altitude, the speed continues to decline steadily until reaching 0 km/s at ground level. Concurrently, as the angle of attack decreases to 15°, there is a corresponding reduction in the values.

Given the relatively small surface area of the vertical fin and the length of the actuator, it is expected to have a power requirement in the range of a couple of hundred watts.

Comparatively, the maximum power demand for an Electro-Hydrostatic Actuator (EHA) with overall efficiency $\eta = 0.67$ stands at 465 W, while for an Electro-Mechanical Actuator (EMA) with overall efficiency $\eta = 0.84$, it is 364 W.

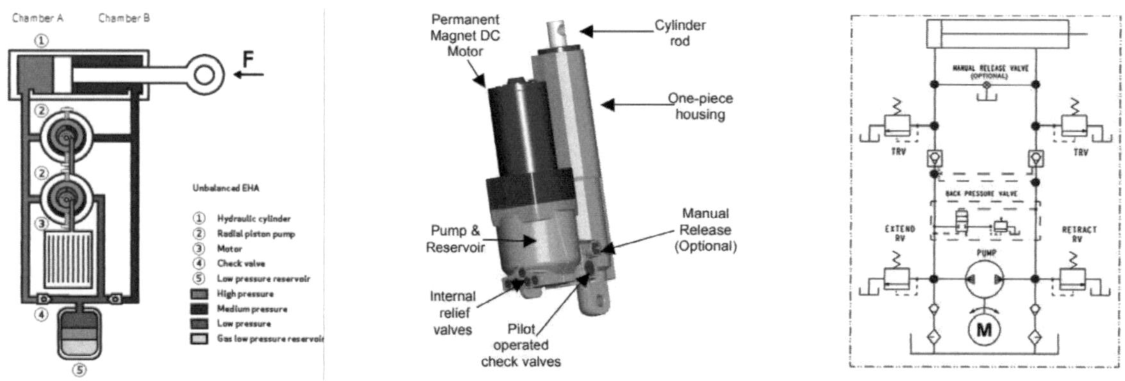

Electro-Hydrostatic Actuators (EHA) proposed by Parker

Specifications for "The Dream Chaser" Vertical Fin (Rudder) EHA

Limit load	6102.4 N
Length of the EHA	126.7 mm
Stroke	38.3 mm
Rod diameter	3 mm
Piston diameter	17.07 mm
Angular rate	60°/s
Power required	465 W
Mass	1.769 kg
System pressure	5,000 psi

Narrow-Body Aircraft e-Taxi System
Battery Modules Vs. Hydrogen Fuel Cell

Ilan Berlowitz
IBAero Engineering
iberlowitz24@gmail.com

Abstract

The concept of taxiing aircraft without relying on main engine thrust is not new. As far back as 1977, Aerospatiale's design office proposed the idea of "motorized wheels for autonomous taxiing" for a subsonic aircraft weighing 76 tons. However, it is only in recent years, with advancements in technology and the rise in oil prices, that this idea has become increasingly appealing as a means to reduce fuel consumption and emissions.

Traditionally, aircraft taxiing has relied on engine thrust, with one or both engines engaged, depending on operational policies. However, the surge in passenger numbers and flight frequencies has led to congestion and increased emissions at airports. To address these challenges, the aviation industry is actively seeking ways to reduce emissions and noise pollution during ground operations, such as taxiing, while also streamlining turnaround times.

In alignment with agreement ACARE 2050, which aim to decarbonize aviation, ambitious targets have been set to achieve zero-emission airport operations by 2050, including emissions-free taxiing.

Currently, TaxiBot stands as the only operational and certified aircraft taxiing system that can manoeuvre an aircraft to and from the runway without using the main engines. TaxiBot serves as an aircraft tractor. Notably, pilots can control TaxiBot using standard controls through a Nose Landing Gear interface, eliminating the need for external steering.

Another emerging solution is WheelTug, designed to enable aircraft to taxi without external towing vehicles or engine thrust. Utilizing electric motors installed on the nose landing gear wheels, WheelTug facilitates forward and backward taxiing, powered by the aircraft's Auxiliary Power Unit (APU).

While both systems offer autonomous taxiing capabilities, they differ significantly in their approach. TaxiBot is a Ground Support Equipment (GSE), whereas WheelTug operates as an onboard system.

TaxiBot towing a Lufthansa aircraft (left), WheelTug e-Taxi system (center), Electric Green Taxiing System EGTS main landing gear motor (right)

The e-Taxi system offers:

- Taxi-out and taxi-in with all engines stopped capability,
- A total aircraft autonomy allowing the aircraft to "pushback" without any tractor.

Introducing the e-Taxi system, which enables taxiing with all engines off, provides complete aircraft autonomy, and eliminates the need for a tow tractor. As an onboard solution, e-Taxi requires substantial hardware installation, adding weight to the aircraft. Consequently, it is primarily considered for short and medium-range aircraft, such as the B737 and A320 Family fleet.

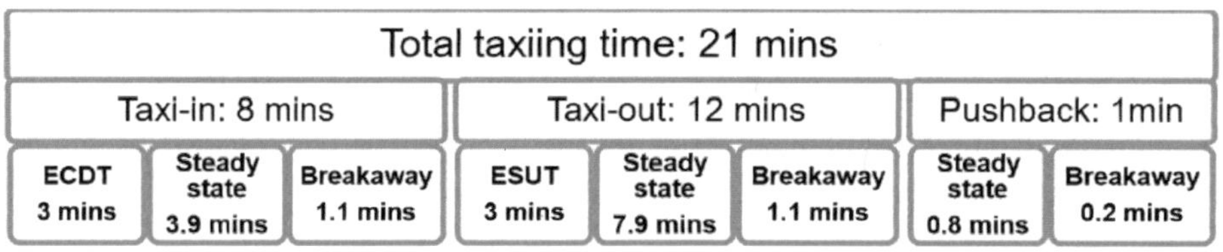

Taxiing Time Allocation

The performance targets of the system are detailed in the table below:

Performance Targets of the e-Taxi System

Phase	**Targets**		**Required Energy kWh**
Pushback	Speed	3.24 knots at MTOW	0.5
	Acceleration	$0.18 m/s^2$	
Taxi-out	Speed	23 knots at MTOW	29
	Acceleration	$0.18 m/s^2$	
Taxi-in	Speed	23 knots at MLW	14
	Acceleration	$0.18 m/s^2$	
Total Energy Required			**43.5**

MTOW - Maximum Take-Off Weight
MLW - Maximum Landing Weight

Continuing the trend of green taxiing initiatives, this study explores the feasibility of implementing a battery-powered landing gear taxiing system based on "Tesla" battery modules specifications. Detailed design considerations include the battery modules, motor, gear, battery management system (BMS), cooling system, and cockpit controller, along with reliability and safety analysis.

The figure below describes the energy densities of few battery modules:

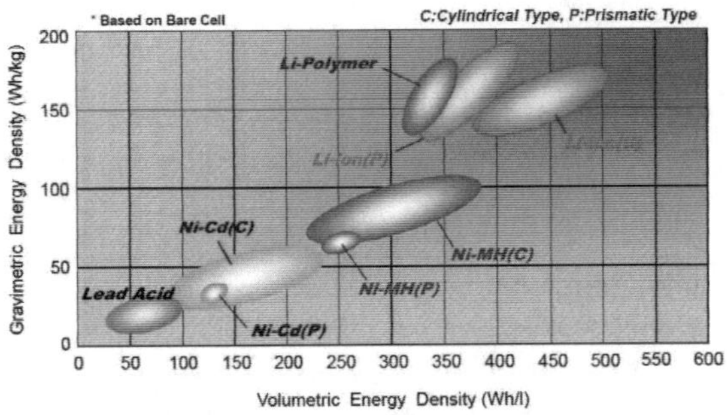

Volumetric and Gravimetric Energy Densities of Battery Modules

"Tesla" Battery Modules

Similar to the Electric Green Taxiing System (EGTS), this system utilizes electric motors on the main landing gear wheels powered by battery modules. The modules, comprised of eight battery modules connected in series, delivers the required voltage and power output for taxiing operations. To ensure safety and reliability, active sensors, detectors, fuses, and circuit breakers are integrated into the system.

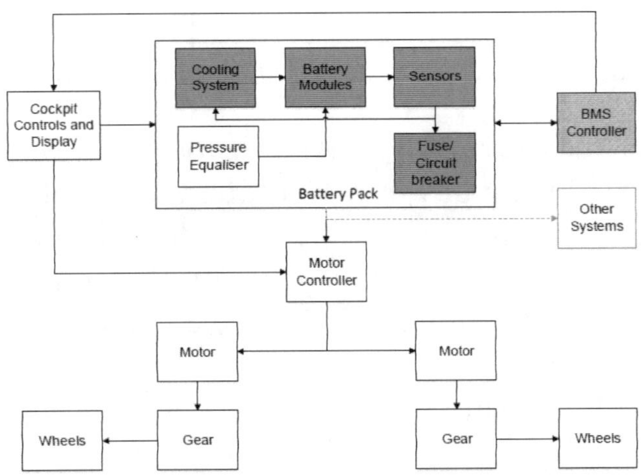

Battery Modules e-Taxi System Architecture

The battery modules with dimensions 1,410 mm · 430 mm · 110 mm and associated BMS controller have been placed into an LD3-45 container to be fitted into the cargo hold of the aircraft and hence can be replaced and effectively recharged at aircraft hangars. It is easy for maintenance personnel to install, handle and carry out repairs by taking out the container and hence the system.

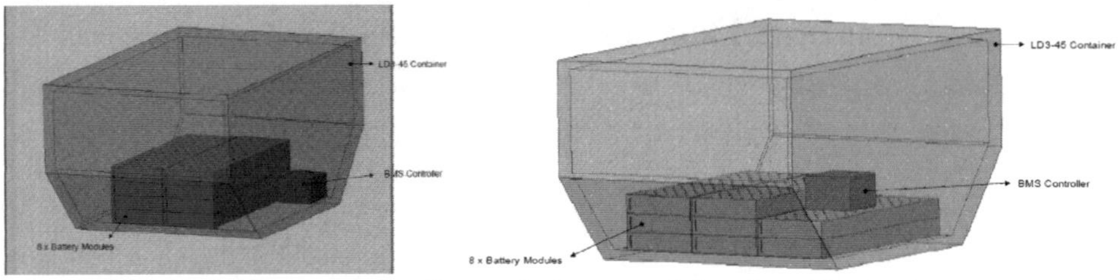

Arrangement of Modules Inside the Container

The system disconnects from the motors and switches off the power supply as soon as the taxi-out stage or taxi-in stage is complete so that it does not work and remains isolated from the aircraft electrical system.

The total weight of the Battery Modules e-Taxi System is 485 kg.

A separate study conducts an analysis of the initial sizing and high-level design of a hydrogen fuel cell-powered e-Taxi system, augmented by regenerative braking. This study aims to determine the potential of such a system in effectively reducing CO_2 emissions in Boeing 737-800 aircraft operations.

Comparison of Energy Carriers

Energy carrier	**Gravimetric energy density (kWh/Kg)**	**Volumetric energy density (kWh/L)**
Jet Fuel	12	10.4
Batteries	0.3	0.8
Liquid Hydrogen	33	2.4

The results of the case study reveal that the utilization of a planetary gearbox with a gear ratio of 13:1 requires peak power and torque ratings of 297 kW and 900 Nm, respectively, for the electric motors driving the left and right Main Landing Gear (MLG). Furthermore, the integration of regenerative braking results in an average energy saving of 7.5% across simulated drive cycle profiles.

Hydrogen is stored within a spherical pressure vessel designed to accommodate sufficient fuel for eight flight legs. Additionally, a 5-kWh lithium-ion battery is selected for storing regenerated energy.

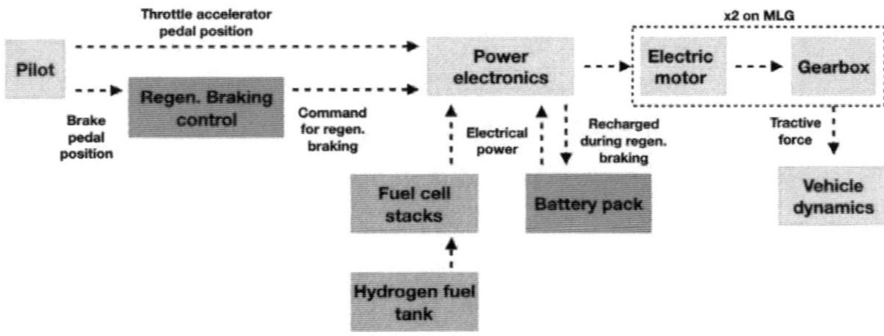

Schematic Overview of the e-Taxi System

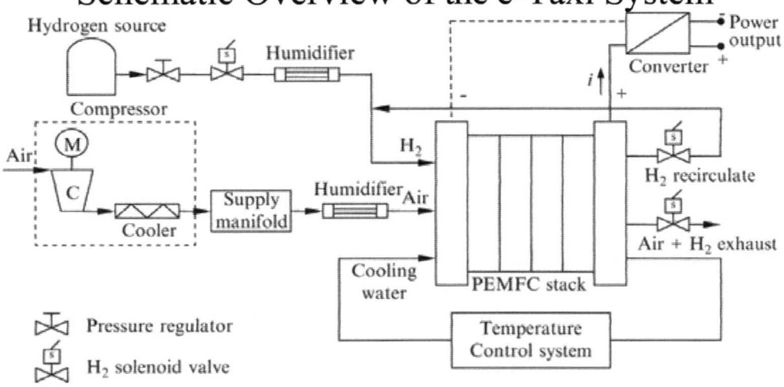

Schematic Overview of the Hydrogen Fuel Cell

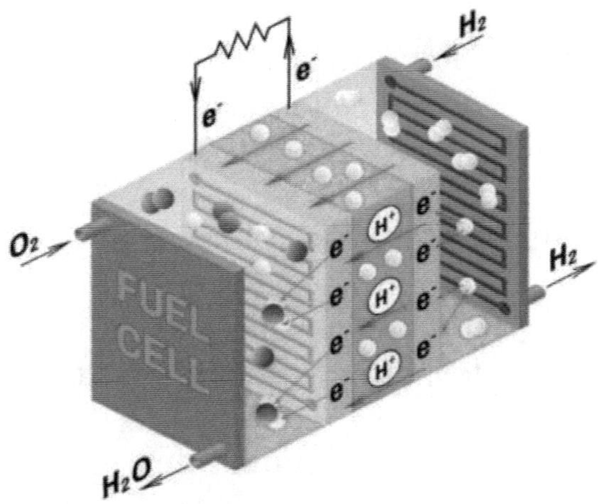

Polymer Electrolyte Membrane Fuel Cell (PEMFC)

The total weight of the Hydrogen Fuel Cell e-Taxi System is 975 kg.

Two key topics to gain a comprehensive understanding of the system:

- Determining the additional hydrogen required to power all subsystems in an engines-off condition, potentially impacting system mass and the frequency of hydrogen refueling.
- Investigating hydrogen supply infrastructure, refuelling procedures, and times at airports, particularly for outstation airports.

Transport Aircraft Cabin Air Quality Improvement

Ilan Berlowitz
IBAero Engineering
iberlowitz24@gmail.com

Abstract

In recent years, there have been increasing instances of aircraft crew members being compelled to retire due to chronic health issues linked to poor cabin air quality. The term 'aerotoxic syndrome' emerged in 2000 amidst a surge of reports detailing such ailments. Some of these health problems have been attributed to the presence of toxic chemicals circulating in the occupied areas of aircraft, potentially stemming from leaks of lubricant oils. One such compound found in synthetic oils is tricresyl phosphate (TCP), known to be harmful to humans in large doses, prompting significant research attention.

A variety of sources were consulted to assess the current evidence on cabin air quality and 'aerotoxic syndrome'. However, the stance of commercial aircraft OEMs has often been dismissive, asserting that the exchanged air, coupled with HEPA filters, ensures air quality in aircraft cabins is on par with or superior to that of general building environments.

At the International Aircraft Cabin Air Conference in 2023, it was noted that the issue of 'aerotoxic syndrome' was recognized as early as the 1950s. Despite over 150 research studies and initiatives to date, it remains a recognized flight safety concern. Both the FAA and EASA have been criticized for their failure to adequately address this flight safety issue and safeguard the health of crew and passengers. Legal actions against airlines and aircraft manufacturers are on the rise, underscoring the urgency of implementing effective total cabin air filtration systems and onboard sensors to enhance flight safety and protect public health.

The breathing air is provided to passengers and crews unfiltered directly from the compression section of the engines or from the Auxiliary Power Unit (APU). This is a process known as 'bleed air' because it is 'bled' from the hot compression section of the engine. The 'bleed air' is not filtered and is known to be contaminated with synthetic jet engine oils and hydraulic fluids.

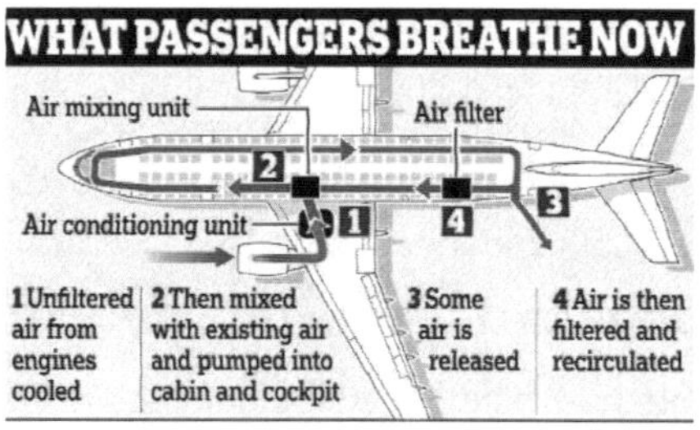

The industry filters the 'bleed air' used for the Fuel Tank Inerting System (FTIS). FTIS was introduced after the TWA 800 tragedy to prevent a fuel tank ignition. The FITS system works by providing a nitrogen rich environment in the fuel tank. The system also uses bleed air, but because of the presence of engine oil fumes in the 'bleed air' and their adverse effects on the system, this 'bleed air' is filtered. **Why does the industry not also filter the 'bleed air' people are breathing?**

A case study utilizing an A320 ECS was conducted to determine the optimal placement of filters, considering effectiveness, air flow pressure drop, and maintenance ease. The study identified two activated carbon filters downstream of the air conditioning packs, along with a bleed air filter/VOC catalytic converter single element package downstream of the engines/auxiliary power unit (APU) regulator and pre-cooler, as the most effective solution. Additionally, the option to incorporate MEMS sensors either in the aircraft cabin, flight deck, or near the bleed air tapping points of the engines and APU was suggested.

PTI Technologies Inc., developed a bleed air filtration solution between the bleed air regulator/precooler and the ECS pack, combined with recirculation HEPA filters, aimed at improving cabin air quality.

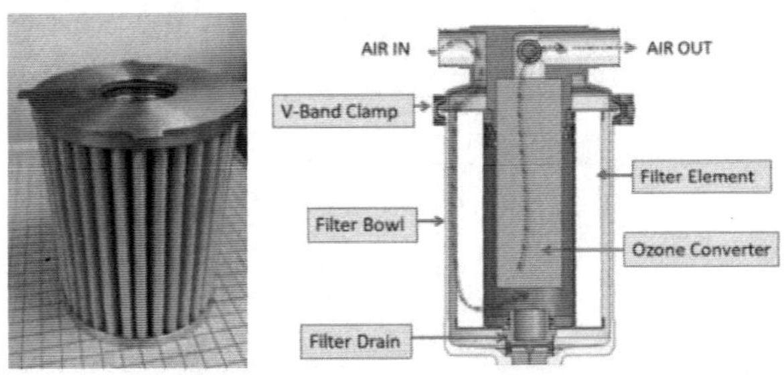

PTI Technologies Inc., New Patented Bleed Air Filter/VOC Catalytic Converter Single Element Package

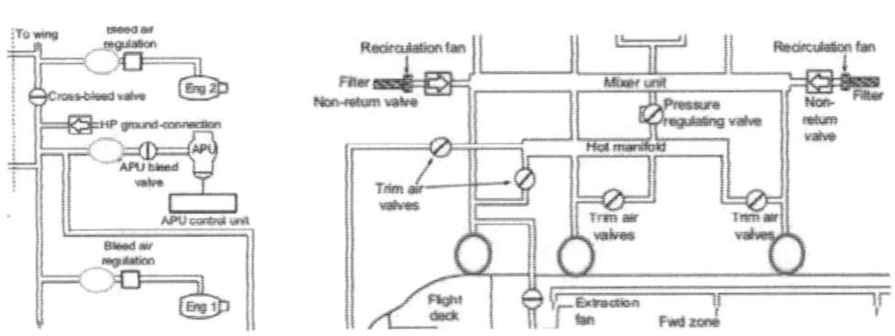

A320 ECS Possible Location of Sensors

The world's first real-time, poisonous compound sensor for aircraft cabins has been verified by UK's National Physical Laboratory so that pilots and cabin crew will soon be able to know if and when poisonous compounds are present in the aircraft cabin, and take appropriate action for the health and safety of themselves and their passengers. **VN Aerotoxic Detection Solutions Ltd, May 2022**

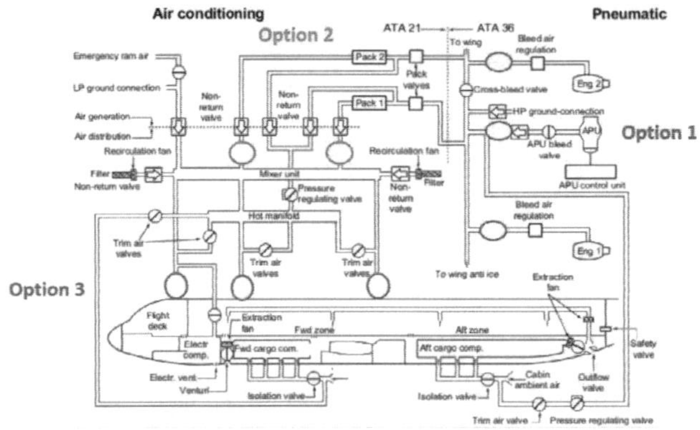

A320 ECS Possible Location of Filters

Production & Freighter Conversions

Traditional Design Approach:

- Affect Boeing and Airbus production & freighter conversions.
- Deleted all passenger amenities including individual gasper system, lavatory & galley ventilation system.
- Delete or partially removed the recirculation fan system & HEPA filters and flight deck associated control logic and switches → reduce operational empty weight (OEW) and ease "C Check" maintenance procedures.
- Air (bleed) from outside only → fuel consumption, contains contamination.
- Increase cabin & flight deck ventilation rate from NORM to HIGH → increase fuel burn.

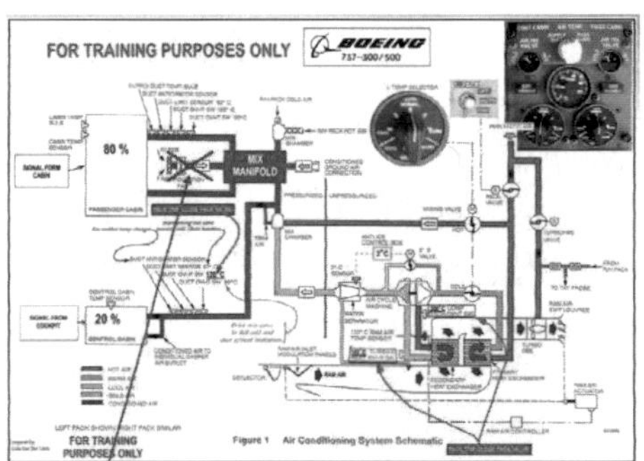

Recirculation fan & HEPA filter and flight deck control logic and switches are removed

Proposed Design Approach:

For production & freighter conversions with multiple cabin zones, such as flight deck, supernumeraries' compartment, rest areas, and forward & aft main cargo compartments, it is recommended to retain the recirculation fan system and HEPA filters. This approach reduces conversion tasks, increases fuel efficiency by 0.8%, and decreases contamination. Setting the PACK flow to NORM or HIGH ensures same efficient air exchange rate for recirculation fans ON (when HEPA filters are installed). Combined with PTI Technologies patented bleed filter/VOC catalytic converter single element package → **better cabin air quality**.

A data-driven strategy for minimizing parts count in aircraft design

Yuval Freed, Maoz Koren
Aviation Group, Israel Aerospace Industries, Ben Gurion International Airport, 70100, Israel

Minimizing parts in aircraft design is of significant importance for several reasons. Firstly, a design with fewer components reduces the design and substantiation efforts during the development phase. Furthermore, a reduced number of parts simplifies the manufacturing process and accompanied logistic footprint, leading to lower production costs and faster assembly times. Finally, a simplified design with fewer parts tends to be more reliable and easier to maintain. This not only translates to reduced maintenance costs but also improves the aircraft's overall availability by minimizing downtime for repairs or retrofits. The importance of minimizing parts in aircraft design is thus linked to achieving a delicate balance between performance optimization, maintenance efficiency, and economic considerations in the dynamic and complex field of the aviation industry.

Several researchers studied how standardization of parts or components is highly beneficial in manufacturing processes [1-4]. Boeing even proposed a platform for an overall aircraft design in which new models may be rapidly designed to fit specific marketplace niches by rescaling existing parts or group of parts [5]. While industry standardization can be easily achieved for development of new products, it is highly challenging and time consuming to employ such approach to already certified products.

The integration of machine learning strategies in the structural design and analysis of airframes marks a transformative leap in the aviation industry. Machine learning algorithms can process immense volumes of data related to aerodynamics, materials science, and structural integrity, providing unprecedented insights into optimal airframe configurations. These algorithms excel at recognizing intricate patterns, enabling more efficient designs that minimize weight while maximizing strength [6]. Additionally, machine learning facilitates predictive analysis of potential stress points and vulnerabilities, allowing for preemptive adjustments during the design phase to enhance durability and safety [7]. This data-driven approach not only expedites the design process but also contributes to the development of more resilient and high-performance airframes. As the aviation sector continues to evolve, the utilization of machine learning in structural design promises to revolutionize the way aircraft are conceived, offering a harmonious blend of efficiency, safety, and innovation.

In this investigation, a machine learning methodology is applied to standardize components through a hierarchical clustering process [8]. Using structural splices in a business jet's fuselage sections as a benchmark problem, we identified hundreds of distinct aluminum parts distinguished by dimensions (length, width, thickness), number of fasteners, and shape. The hierarchical clustering procedure was employed to reduce the number of parts. The results of this process are illustrated in Figure 1, where the lower level of the clustering process depicts hundreds of individual parts. As we ascend the levels, broader categories emerge, representing families of parts that share similar characteristics.

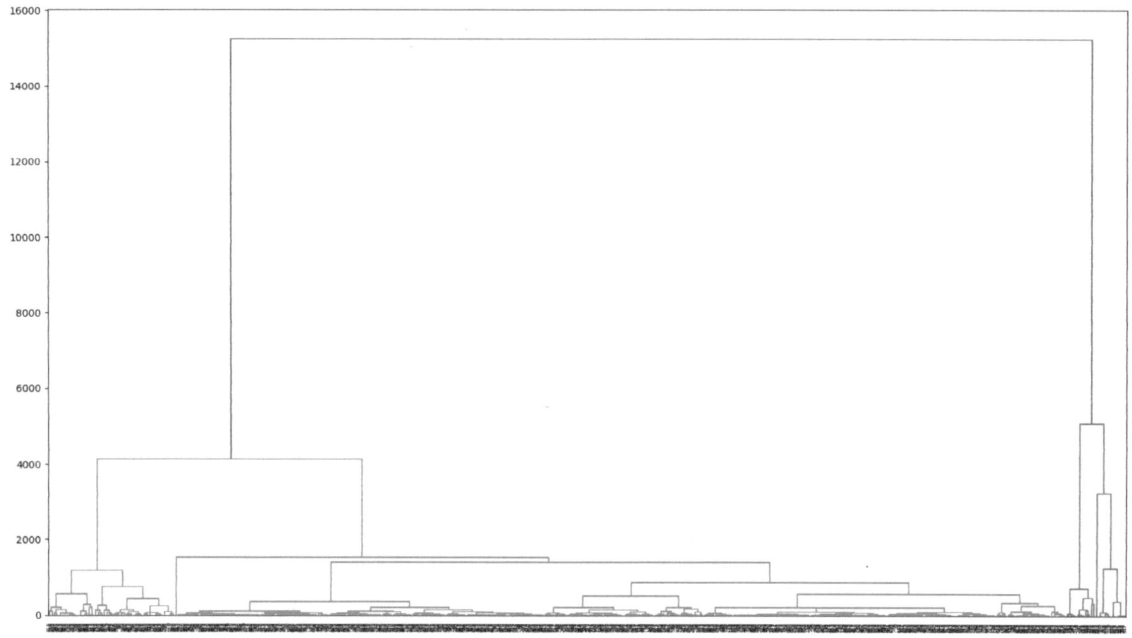

Figure 1: Flow-chart describing hierarchical clustering for splices in a business jet fuselage. The base level of chart represents ~ 500 parts, and each level up to the top represents clustering to fewer families

Next, the effectiveness of the clustering process was assessed in relation to the additional weight attributed to consolidated parts and the reduction in the overall number of components. As anticipated, there is a rise in the weight penalty for a smaller number of clusters. Nonetheless, as depicted in Figure 2, when dealing with approximately 100 parts, the weight penalty is about 10 kg, a reasonable figure that exerts minimal impact on the aircraft's performance. This reduction translates to an 80% decrease in the number of parts, resulting in a substantial reduction in associated manufacturing and logistical costs.

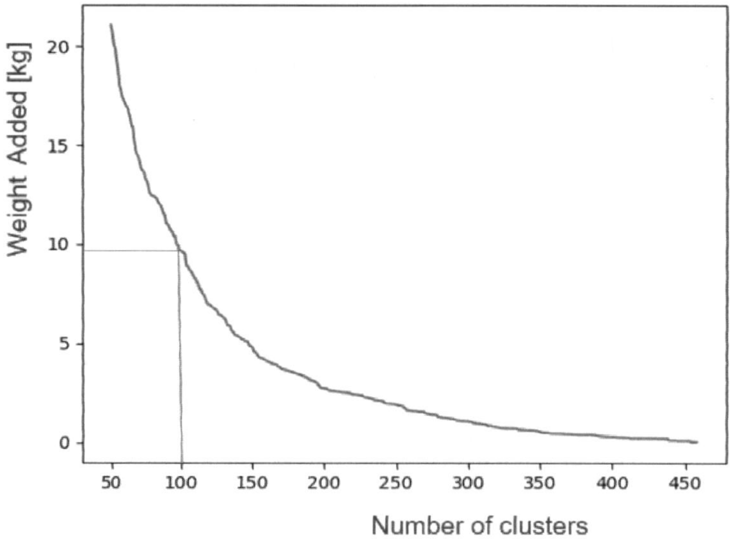

Figure 2: Weight penalty as a function of the parts clustering

References:

[1] Collier DA. The measurement and operating benefits of component part commonality. Decision Sciences. 1981 Jan;12(1):85-96.

[2] Collier DA. Aggregate safety stock levels and component part commonality. Management science. 1982 Nov;28(11):1296-303.

[3] Tsubone H, Matsuura H, Satoh S. Component part commonality and process flexibility effects on manufacturing performance. THE INTERNATIONAL JOURNAL OF PRODUCTION RESEARCH. 1994 Oct 1;32(10):2479-93.

[4] Perera HS, Nagarur N, Tabucanon MT. Component part standardization: A way to reduce the life-cycle costs of products. International journal of production economics. 1999 Apr 20;60:109-16.

[5] Meyera MH, DeToreb A. Perspective: creating a platform-based approach for developing new services. Journal of Product Innovation Management: AN International Publication of the Product Development & Management Association. 2001 May;18(3):188-204.

[6] Freed Y., Dorfman B. and M. Koren, Aircraft Loads Assessment and its Effect on Aircraft Structure: Machine Learning Approach, 62nd Israel Annual conference on Aerospace Sciences, 2023, Tel Aviv, Israel

[7] Pinhas E., Freed Y., Buimovich Y., Dolev O., Noivirt G. and Ofir Y., Anomalies Detection During Full Scale Fatigue Test Using Machine Learning Approaches, 62nd Israel Annual conference on Aerospace Sciences, 2023, Tel Aviv, Israel

[8] Ward Jr JH. Hierarchical grouping to optimize an objective function. Journal of the American Statistical Association. 1963 Mar 1;58(301):236-44.

Real-Time Health Monitoring of Aeronautical Structures via Sensitivity Tests Utilizing Principal Component Analysis

Y. Ofir[1], U. Ben-Simon[1], S. Shoham[1], I. Kressel[1]
J. Bohbot[2], M. Tur[2]
[1]IAI Aviation Group, Ben Gurion International Airport, Israel
[2]School of Electrical Engineering, Tel-Aviv University, Tel-Aviv Israel

Modern airborne structures, subjected to high maneuverability and exposed to demanding launch and landing conditions, necessitate continuous monitoring of their structural integrity and airworthiness. Structural Health Monitoring (SHM) proves particularly valuable in the case of composite aerostructures, where traditional inspection methods for critical components face challenges due to limited accessibility. An effective online SHM system can be achieved by employing fiber optic sensors, specifically Fiber Bragg Gratings (FBGs). These sensors facilitate the integration of a 'fiber optic nervous system' into the composite structure, enabling the detection, measurement, and communication of various environmental and structural parameters.

The success of an SHM system for aircraft structures relies not only on selecting the right sensors and making precise measurements but also on the processing of sensor data to derive critical information regarding operational loads and potential structural damage. Such a system allows for continuous monitoring of the structural health of these aerostructures, ensuring real-time (few seconds) assessment of their airworthiness and enabling prompt corrective actions if necessary, without the need for unnecessary grounding downtime. This holds significant strategic importance, enhancing fleet readiness and safety considerations.

Principal Component Analysis (PCA) is a well-established technique in multivariate analysis. It was first introduced by Pearson (1901), and developed independently by Hotelling (1933). PCA gained widespread recognition with the proliferation of high computing capabilities readily available in off-the-shelf computers. The core concept behind PCA is to reduce the dimensionality of a dataset containing numerous correlated variables, while preserving the majority of the data's variability. This reduction results in substantial computational efficiency without significant information loss.

The process involves transforming the data into a new set of variables called principal components, which are uncorrelated and ordered in a way that the first few components capture most of the original data variation. Computing these principal components simplifies to solving an eigenvalue-eigenvector problem for a positive-semidefinite symmetric matrix.

In the field of Structural Health Monitoring (SHM), PCA has found extensive applications. It has been employed for various purposes, including the analysis of vibration signals to reduce dimensionality, eliminating environmental effects from vibration characteristics, extracting features related to structural damage, distinguishing features between damaged and undamaged structures, and classifying acoustic emission transients. There are PCA-based tools available that provide insights into the model and its characteristics.

We utilize two measures, the Hotelling's T-squared distribution (T^2) and the Q-statistic, as indicators of faults. These statistics have been demonstrated to exhibit high sensitivity to structural anomalies, even at small scales. When integrated into a specially designed framework, the PCA model and these statistics form a decision support tool that offers dependable real-time information regarding the airborne structure. Such information can prove invaluable for maintenance, assessing the structure's lifespan, and making operational decisions related to mission continuity.

We demonstrate the performance of the method on real flight structural data (strains). In this case, the PCA model is based on regular flight envelope that includes standard maneuvers. The system successfully identifies a harsh maneuver that deviates from the pre-defined confidence limits. See Figure 1.

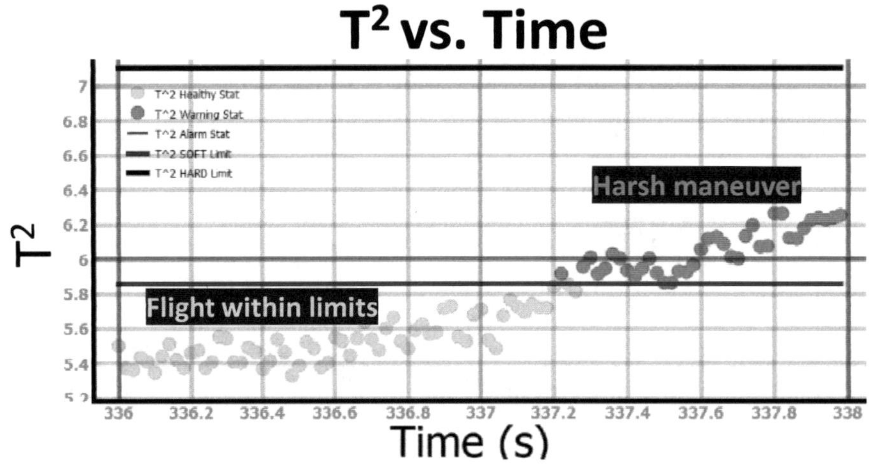

Figure 1 - Deviation from flight envelope indicated by the Hotelling T-squared statistics

Numerical Prediction of stringer de-bonding in composite stiffened panels subjected to combined loading

D. Bardenstein, A. Lukatsky and I. Kressel
Israel Aerospace Industries, Ben Gurion International Airport Israel

B. M. Smyers, P. M. Knoth and S. B. Clay
USAF Research Laboratory, Wright-Patterson Air Force Base, USA

In this work "Fail-Safe Technology for Bonded Unitized Composite Structures" is developing and validating new concept that enable the use of slow-damage growth design and certification for unitized composite aircraft structures. The Israel Aerospace Industries, in collaboration with the US Air Force Research Laboratory, designed, analyzed, built, and tested a novel structural concept that is based on a "modified-stiffness" concept, in which mushroom-shaped features are incorporated in the stringers of the stiffened panels in order to retard the growth of skin-stiffener de-bonds. This work focused mainly in the development, calibration and validation of a static F.E analysis approach to predict damage initiation, damage growth and damage arrestment .

Figure 1 presents a MDAF and a Reference Composite stiffened panel specimen, where an initial de-bond was implemented for both. The skin edges were simply-supported and at the 1st' stage, a compression loading was applied such to cause 3000us in the skin. At the 2nd' stage, the compression load was maintained and a Pull loading was applied until a full stringer de-bonding occured.

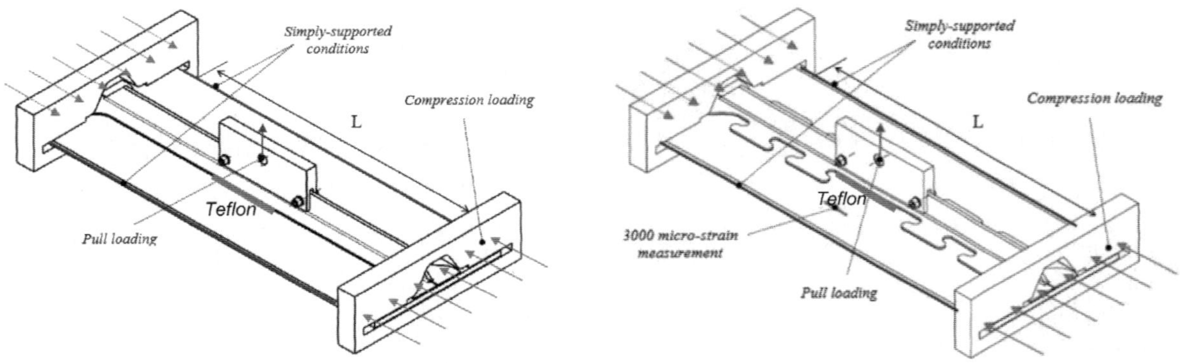

Figure 1: *MDAF* and Reference composite stiffened panel specimens

A *CZM* numerical F.E (*MSC NASTRAN/PATRAN*) analysis was performed in order to predict the *MDAF* and reference panels response to combine compression and Pull loading - the Load/Displacement curves for both panels are presented by the blue curves in Figure 2. It can be seen that the mushrooms may actually retard the growth of skin-stiffener de-bond and once the initial de-bond damage reaches the first mushroom front set, it will be retarded or stopped, and only after an additional significant loading event, it may overcome the mushroom bond-line and propagate further to the next mushroom. On the other hand, the reference panel exhibited an uncontrolled damage propagation and once the critical load was reached, the damage propagated until a full skin and stringer separation occurred. In addition, it can be seen that the *MDAF* concept has much higher "toughness rate" in comparison to the reference concept and that there is approximately 5% increase in the *MDAF* specimen's critical load.

The red curves in Figure 2 presents the mechanical experimental results. It can be seen that while a good correlation was obtained for the maximum load criteria, the F.E model did not predict well the load drop for the *MDAF* specimen. This can be explained by the fact that in the numerical analysis, the damage may propagate through the adhesive interface only, while in the experiments, a mixed interlaminar mode damage propagation of the adhesive interface and first skin ply took place. Finally, as was predicted in the numerical analysis, much higher "toughness rate" was demonstrated for the *MDAF* specimen.

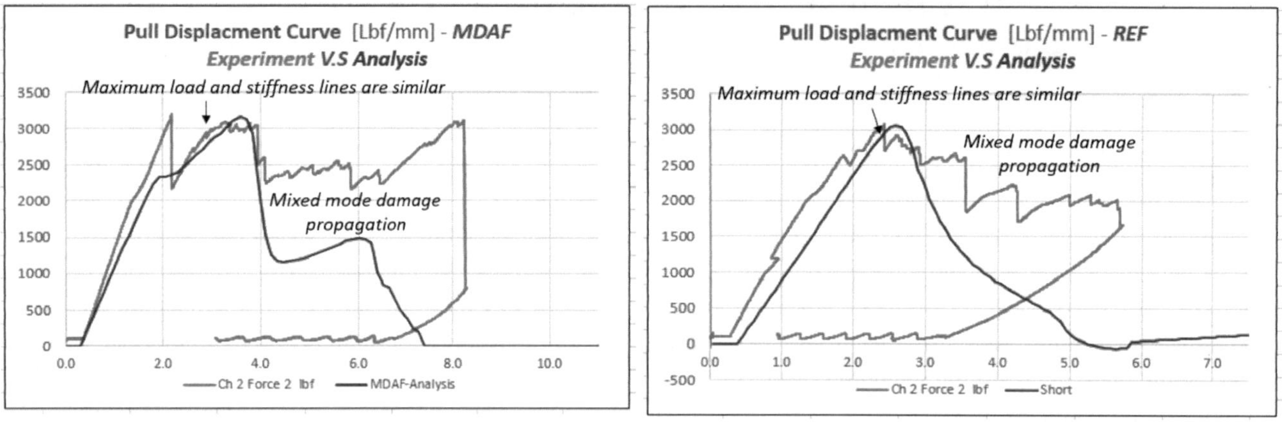

Figure 2: Experimental V.S. Analysis Load/Displacement curves for the MDAF and Reference Panels

An Open Source Quadcopter Platform for Simulink

Joseph Attias, Yael Marciano, Ruslan Arhipov and Daniel Zelazo
Technion-Israel Institute of Technology

1 Abstract

The Faculty of Aerospace Engineering at the Technion offers an advanced control lab course to undergraduates (Control Lab 085705). The course teaches students how to design autonomous flight control and estimation algorithms for a quadcopter. The current course is structured around the "Parrot Minidrones Support Toolbox" from Simulink. Recently, the Parrot drones and their components were taken off the market which makes them obsolete. This project's goal is to design a new drone platform that can be substituted directly into the course and used in quadcopter controller research. The drone should be based on open-source firmware, interchangeable off-the-shelf electronics, and a custom frame that can be integrated with testing hardware. A study was conducted to decide which open-source flight control package would be used as the framework. The study focused on how key requirements for the course would be implemented. These features include integrating the students' Simulink controllers as the primary motor controller, robust two-way communication, and optical flow integration. "Ardupilot" (Arducopter) was chosen among the major open-source flight control packages. Our Simulink wrapper is based on a custom flight mode. The mode pulls raw sensor data from the onboard accelerometer, gyroscope and an external optical-flow module. Additionally, it pulls XYZ estimations from the onboard extended Kalman filter which has been configured to reference the optical-flow measurements. The students interface with the platform through two Simulink models: the *Flight Control System* and the *Dashboard.* Students construct their algorithms in the Flight Control System where they can invoke the firmware compilation and uploading. The Dashboard hosts a display for signal logging and basic flight control. Behind the dashboard is a python script connected by either Wi- Fi or USB to the drone. The current drone is based on a BETAFPV F405 20 Amp AIO Flight Controller. The all-in-one (AIO) combines the flight controller, power distribution, and motor controller (ESC). The flight controller is connected via UART to an ESP Wi-Fi module that has been flashed with support for standard MAVLINK communication. A Micoair MTF-02 optical flow and lidar module is connected via a single UART connection to the flight controller. The drone created in the project fulfils the goals of replacing the Parrot drone system. In doing so, we also deliver a strong framework for implementing Simulink onto an affordable open source quadcopter platform. This feature allows the drone to be used as an accessible testing platform for more advanced research.

RAFAEL
ADVANCED DEFENSE SYSTEMS LTD.
Manor Division
Research and Development

30.4.24

'Smart Wing' – Flight Demonstration of a Novel Composite Wing With Embedded Electronics and Sensing Capabilities

Hay Azulay, Yehonatan Burgansky, Zvika Fayer, Artem Rozman, Taly Sessler Elhassid, Yuval Zait, Amit Kadisheviz, Orian Hazan, Shmulik Suriyano, Ilan Farkash, Erez Zemel

Abstract

The primary advantage of structures made from fiber reinforced composite materials is their improved specific strength. However, one can provide composite structures with additional abilities by transformation of the composite structure to a multi-functional entity. Through this vision, the structure itself embeds within it capabilities beyond the carriage of mechanical loads alone. Such structure multifunctionality in composites was the focus of previous work in Rafael. Current conduction, sensing, communication and actuation applications embedded within the composite structure were successfully demonstrated to name a few. These achievements were made possible through the usage of a novel method for printing a conducting ink on the composite layers and combining it with the composite structure manufacturing process (Fig. 1).

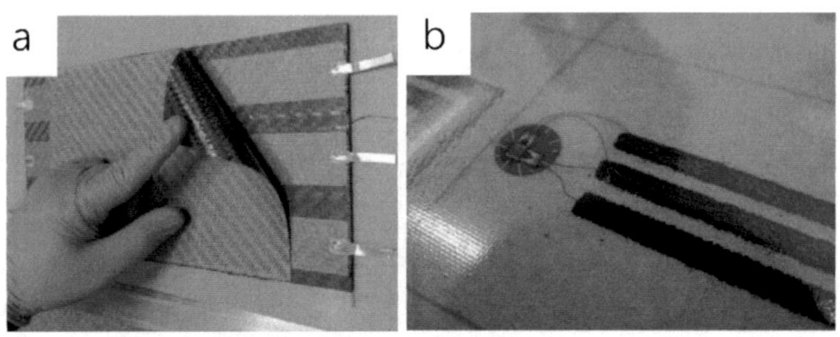

Figure 1: Examples for composite structure multifunctionality through the embedment of electronics between the composite layers. a) Printed conductors between layers during structure manufacturing process, b) embedded sensor within a composite structure

The above mentioned method was implemented for the development of a 'Smart Wing' – a composite wing with embedded electronics. In this specific case, the electrical devices that were demonstrated included a LED and strain gauge sensors, embedded within the composite structure and operated successfully (Fig. 2).

Figure 2: The composite 'Smart Wing' with embedded electronics operating a LED and sensors

Finally, the wing was assembled on a UAV platform and was tested during a flight experiment, demonstrating a successful operation of its multifunctional electrical and sensing embedded features (Fig. 3).

Figure 3: The composite 'Smart Wing' assembled on a UAV platform during a flight test

Signal from the sensors was transmitted to a ground station during the flight. This data enabled the presentation of a 'digital twin' model of the wing on the ground station in real-time, based on the signal the sensors acquire from within the flown wing.

This demonstration of 'smart' composite structures and their proof during flight opens possibilities to embed many different electrical devices within the structure foremost for space and weight reduction and also for the enhancement of the platform capabilities in several different aspects – a topic that is the focus of following research.

Hybrid 3D-1D finite element modeling for elastodynamic bending: Preliminary results

Yana Avni [*], Daniel Rabinovich [†] and Dan Givoli [‡]
Faculty of Aerospace Engineering,
Technion - Israel Institute of Technology, Haifa, 3200003, Israel.

An analysis of complex 3D geometries using the finite element (FE) method often involves fine domain discretization, associated with a large number of degrees of freedom (DOF), that demands high computational resources. A common technique to address these issues is to reduce the high-dimensional (HighD) 3D model to a low-dimensional (LowD) model. The LowD model is obtained by approximating the HighD model using dimensional reduction techniques, e.g., asymptotic analysis or integration through the thickness. A case model for such an approach is the bending analysis of a wing, where the maximal bending and the stress distribution in the wing-fuselage connection are the main objects of interest. The 3D wing global bending can be estimated using a 1D Euler-Bernoulli beam theory, but this approximation cannot demonstrate strain and stress distribution. Analysis of the wing's fix-end structural strength requires a comprehensive 3D model.

Such problems can be solved using a mixed-dimensional FE model incorporating both HighD and LowD sub-models [1]. The fixed wing end and its adjacent area can be modeled with HighD precision, whereas the rest of the wing's span could be simplified as a 1D beam representation. This modeling approach requires an effective coupling method between the two sub-models. One of the techniques for mixed-dimensional junctions is the Panasenko method [1]. In this approach, the primary variables along the HighD-LowD interface are maintained with pointwise continuity, and the reduced variables are estimated based on the physical assumptions. The Panashenko coupling technique was successfully applied in numerous problems, predominantly within static or steady-state conditions [2]. Additionally, Amar and Givoli [3] applied the Panasenko method to time-dependent modeling of the lateral deflection of a 2D-1D hybrid membrane. Efrati and Givoli [4] used this method for the 3D-2D hybrid elastodynamic problem.

This study represents a novel application of this method: the 3D-1D hybrid modeling of elastodynamic time-harmonic bending problems. The method is demonstrated on a model including a massive one-side fixed 3D cube coupled to a hybrid 3D-1D beam-like sub-model. Under a concentrated symmetric load applied on the beam's free end, the cube reaction and the hybrid beam deflection were analyzed and compared to a reference

[*] yana.mayorov@campus.technion.ac.il
[†] aedaniel@technion.ac.il
[‡] givolid@technion.ac.il

full-dimensional 3D model, Figure 1. Variations of this scratch mixed-dimensional model can serve for bending analyses of many structures, such as wing and helicopter rotors.

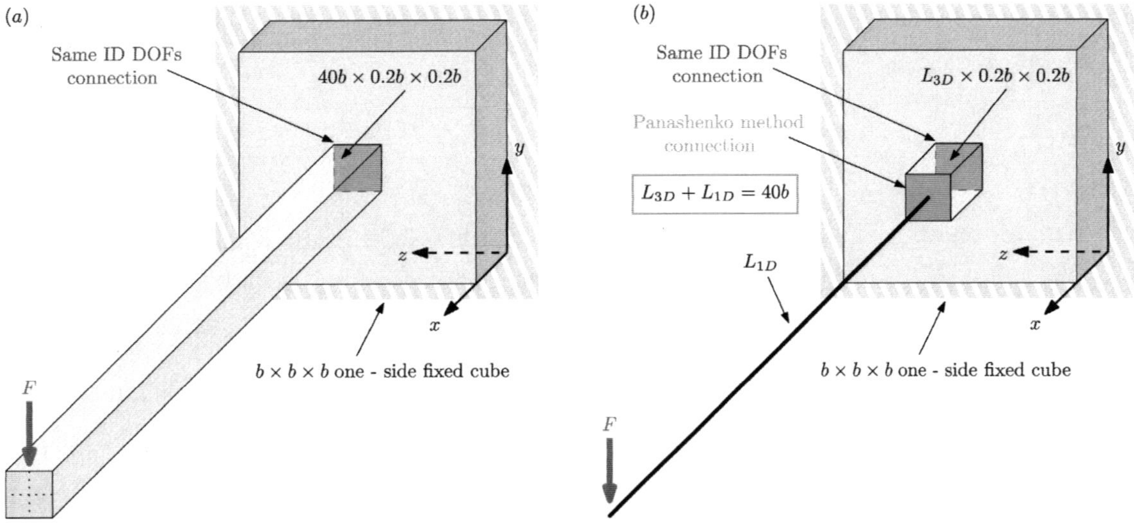

Figure 1. Schematic illustration of the 3D full-dimensional reference model (a) and 3D-1D hybrid model (b) under a bending load.

Implementation of this coupling FE technique is done by characterizing the primary variables on the last 3D interface and the beginning of the 1D beam as the same bending DOF. Additionally, the first derived rotation variable is presumed to be a derivative of the displacement at the 3D solid endpoint, based on the Euler-Bernoulli theory. The hybrid model demonstrates a promising performance in time-harmonic analysis. Furthermore, alongside the nearly perfect correlation between the reference and hybrid model results, the approximated model boasts significantly enhanced computational efficiency compared to the HighD model, owing to a reduction in the DOFs number. The mathematical framework presented here has great potential to simplify complex structural analysis by applying reasonable physics-based approximations.

References

[1] Fontvieille, F., Panasenko, G., and Pousin, J., "FEM implementation for the asymptotic partial decomposition," *Applicable Analysis*, Vol. 86, No. 5, 2007, pp. 519–536.

[2] Panasenko, G., "Method of asymptotic partial decomposition of domain for multi-structures," *Applicable Analysis*, Vol. 96, No. 16, 2017, pp. 2771–2779.

[3] Amar, H. and Givoli, D., "Mixed-dimensional modeling of time-dependent wave problems using the Panasenko construction," *Journal of theoretical and computational acoustics*, Vol. 26, No. 03, 2018, p. 1850034.

[4] Efrati, R. and Givoli, D., "Hybrid 3D-plane finite element modeling for elastodynamics," *Finite Elements in Analysis and Design*, Vol. 210, 2022, p. 103812.

RAFAEL
ADVANCED DEFENSE SYSTEMS LTD.

Manor Division
Research and Development

Abstract: Isogrid Composite Structure From Design to Production

Over a century grid structures were used for various applications. Isogrid is a rib based structure widely used in load bearing parts for aerospace. This structure usually consists of a repeating unit of either triangle, rhombus, pentagon or combination of thereof. Isogrid structure has versatility in design, it is lightweight and has a high stiffness and buckling to weight ratio. However designing an Isogrid structure may be a challenging task due to numerous parameters. For example, geometry, repetitive unit size, rib width, rib thickness and production process have a significant effect on the results. In 21st century, a significant improvement in composite production technology was introduced, that allowed automatic production of more complex structures. One of the most prominent examples is the Automated Fiber Placement (AFP), where the robust, precise and repeatable structure is easily achieved.

In this work the development process of a composite Isogrid structure combining different target functions is presented. The workflow comprises two-scale approach, namely, unit cell model that corresponds to the small scale and the full scale cylindrical model. In small scale the model consists of a skin and a grid structure where various simulation approaches were evaluated. Parameters of grid such as thickness, angle, width, unit size and junction thickness were investigated. In addition, the sensitivity of material properties on the target function were evaluated. Calculation and optimization were performed using MATLAB software and Ansys software. Chosen geometry of Isogrid cylinder was produced using Thermoset AFP, resulting in a lightweight structure. The physical testing on the prototype is undergoing.

POB 2250 Haifa 3102102 Israel | Tel: 972-73-3353504 | Fax: 972-73-3354392 www.rafael.co.il

Matlab Interpolation of Planar Curves with Nonmonotonic Independent Variables

Baruch E. Karlin, Ph.D.
Research Scientist, Haifa, ISRAEL

I. Problem Statement

Data interpolation is used in many fields. Reliable interpolation requires suitable methods and region coverage with sufficient point-density for retaining local shape and features. Independent variables may be nonmonotonic with nonunique values. Curves may be self-crossing and have cusps. Applying Matlab® interpolation functions to such data fails. Web response to *"How to interpolate such data"* queries is *"Impossible."*

II. Methodology

Polyline is a planar piecewise linear curve without gaps. Piecewise linear subdivision of polyline segments adds points but retains the rough shape-approximation inherent in the input data (Figure 1: subdivision offset-polyshape in cyan). Described herein are newly designed PlanB to interpolate problematic data and SimMAD for estimating a quality metric for PlanB testcases, both implemented in Matlab®.

SimMAD estimates the similarity between interpolated and reference curves. It calculates the Median Absolute Deviation of the interpolated points relative to the reference testcase polyline. The reconstruction offset-polyshape combines the reference and the flipped-order interpolated polylines. Matlab® polyshapes are polygons with properties. Similarity metric is the offset-polyshape area divided by the length of the reference polyline. Smaller metric values indicate better fits. Matlab® may issue a polyshape warning.

PlanB imposes monotonicity by using the cumulative length of the input polyline as the independent variable assuming same x and y units. Each polyline component is interpolated with Matlab® MAKIMA for a required number of points. MAKIMA is a Modified Akima Piecewise Cubic Hermite Interpolator that gives good fits to curves with a rapidly varying second derivative (see Matlab® Help). Point-spread along the interpolated curve is determined by the point-spread along the cumulative length of the required polyline. Curve reconstruction quality depends on input point-density. A low-density dataset may belong to distinctly different curves. Extrapolation produces incorrect results and is highly unrecommended.

III. Performance

SimMAD was tested with a 40-points curve of two circular arcs and a 1 unit offset curve. SimMAD estimated the offset with an error of about 1 milli-unit (Figure 2: offset-polyshape in cyan). PlanB was applied to testcases with known parametric curves: spiral, sine, fish, stall, and pentagram (Figures 3÷12). The number of interpolated points was three times the number of input points. For consistency, results are displayed with the designated dependent y-data versus the designated independent x-data regardless of actual dependency.

PlanB testcase reference polyline is a linear piecewise representation of the true curve, and the input data is a rough approximation of that curve. The interpolated point-spread differs from the reference point-spread, but the similarity metric estimated by SimMAD indicates good correspondence (compare Figures 1 and 3). Very sparse input data may cause incorrect curve reconstruction. Adding input points increases output reliability. At least five input points are recommended for sections between large slope changes. Output point-spread is controllable. SimMAD and PlanB work well. Future work: 3D curves and prespecified output x-data.

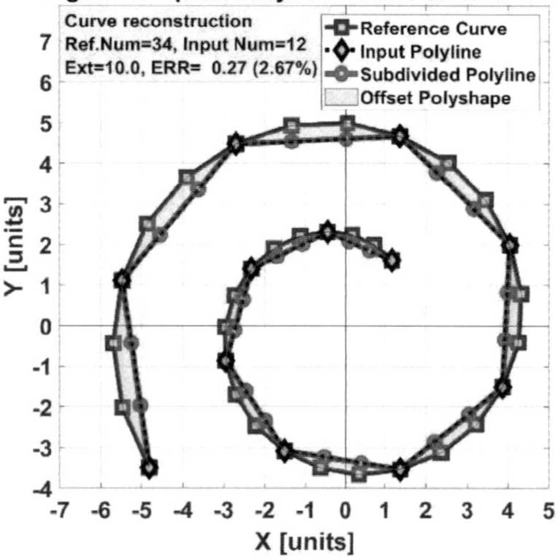

Figure 1: Spiral Polyline Linear Subdivision

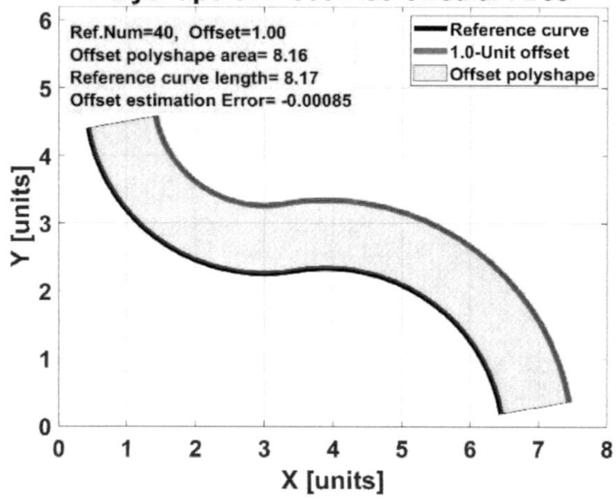

Figure 2: Offset Estimation Using Combined Polyshape of Piecewise Circular Arcs

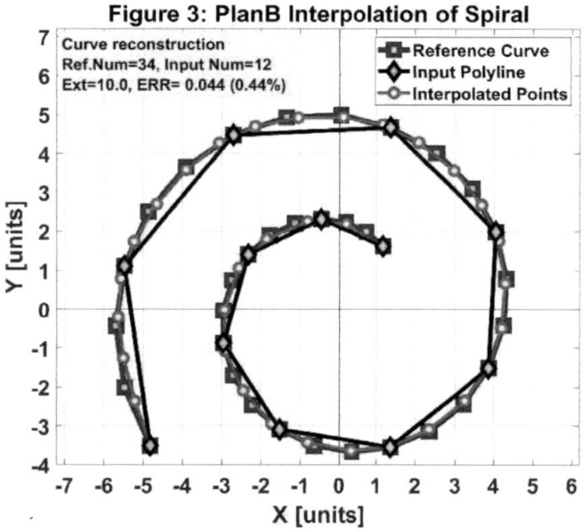

Figure 3: PlanB Interpolation of Spiral

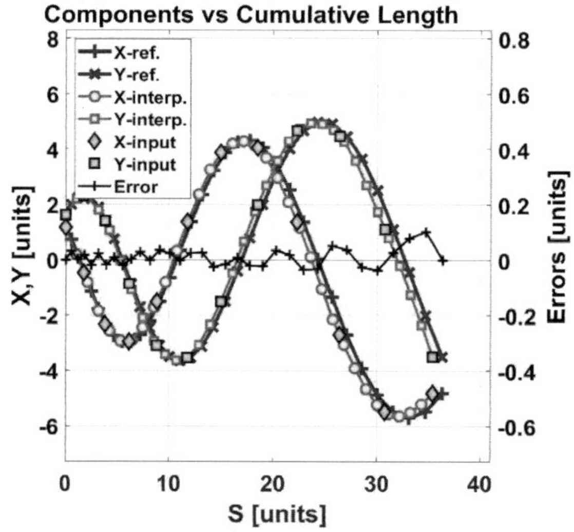

Figure 4: PlanB Interpolation of Spiral Components vs Cumulative Length

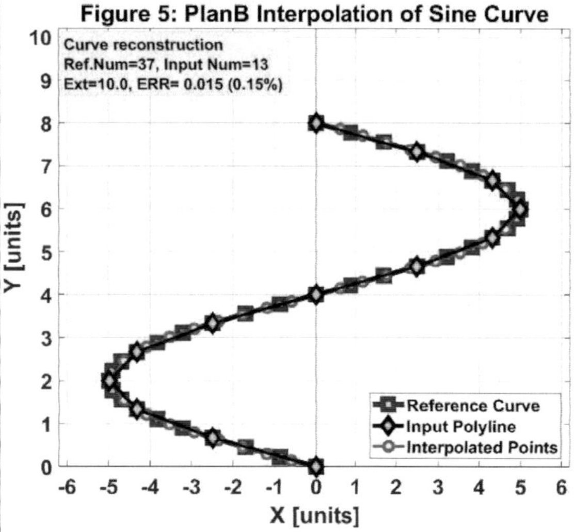

Figure 5: PlanB Interpolation of Sine Curve

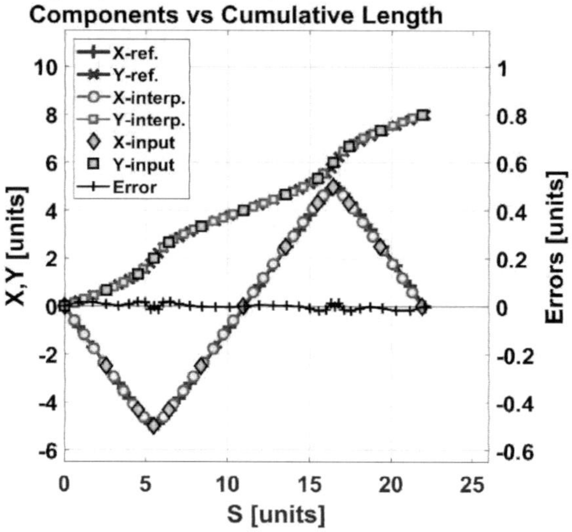

Figure 6: PlanB Interpolation of Sine Curve Components vs Cumulative Length

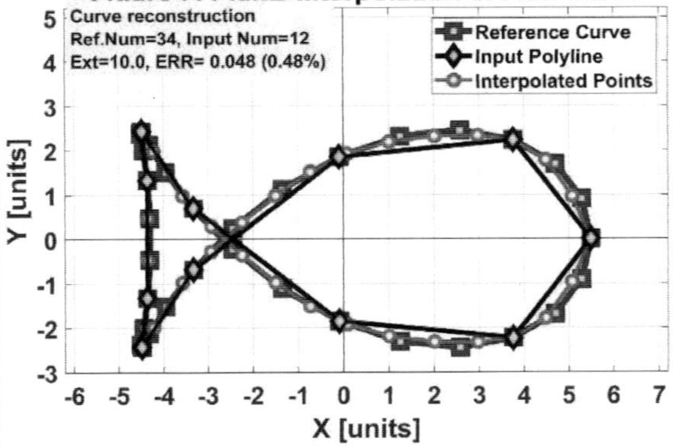

Figure 7: PlanB Interpolation of Fish Curve

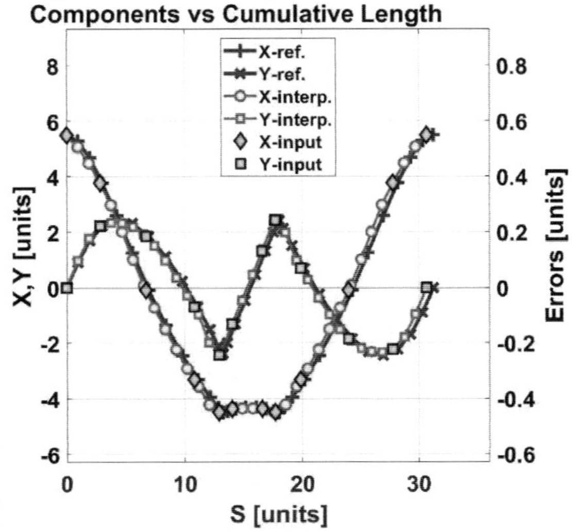

Figure 8: PlanB Interpolation of Fish Curve Components vs Cumulative Length

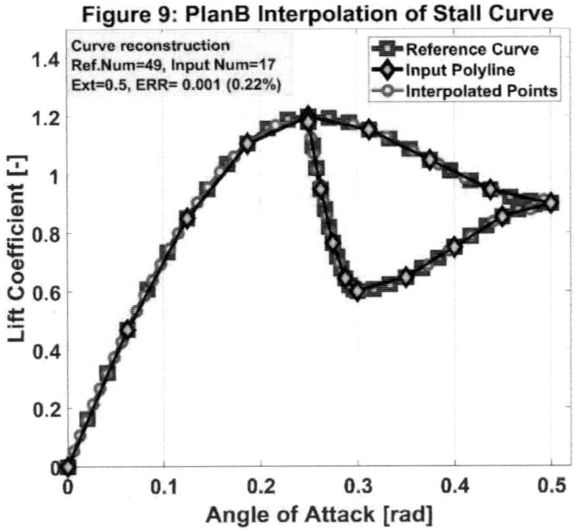

Figure 9: PlanB Interpolation of Stall Curve

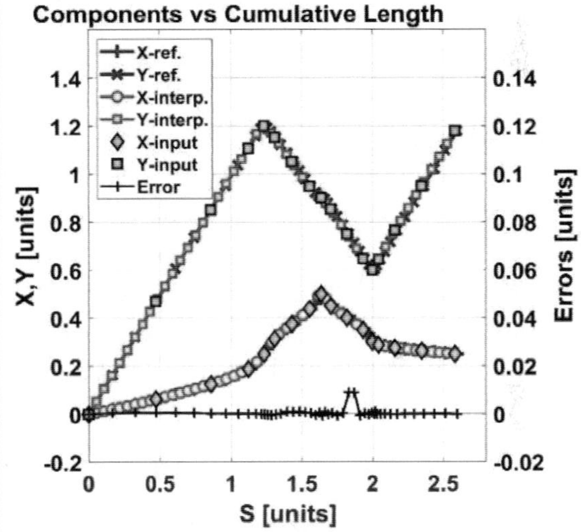

Figure 10: PlanB Interpolation of Stall Curve Components vs Cumulative Length

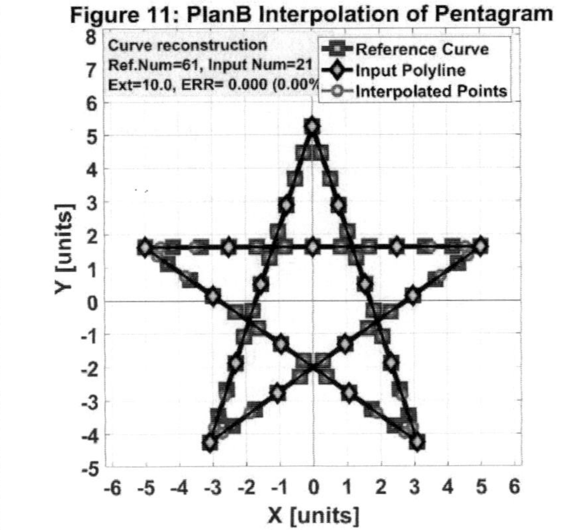

Figure 11: PlanB Interpolation of Pentagram

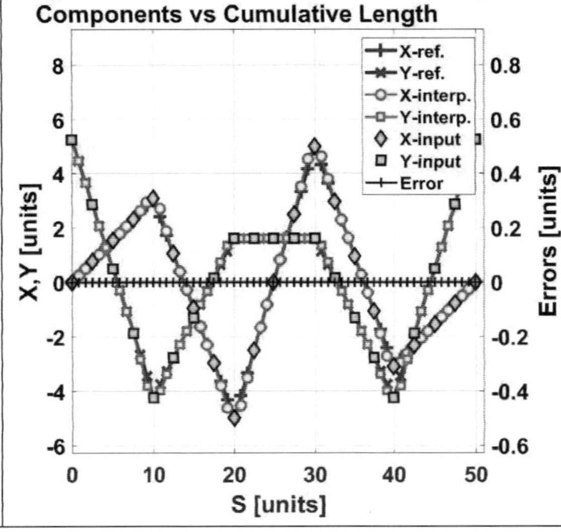

Figure 12: PlanB Interpolation of Pentagram Components vs Cumulative Length

Identification of an Elastic Inclusion Using a Time-Dependent Adjoint Method

Amit Sayag[1], Dan Givoli[2]

[1] Technion-Israel Institute of Technology, Dept. of Aerospace Eng., Haifa 32000, Israel
amitsayag@campus.technion.ac.il

[2] Technion-Israel Institute of Technology, Dept. of Aerospace Eng., Haifa 32000, Israel
givolid@technion.ac.il

Key Words: *Inclusion, Inverse problem, Adjoint method, Gradient, Shape identification, damage identification, Elastodynamics.*

The inverse problem of identifying an inclusion, a small region in the medium where the material properties are significantly different than those of the surroundings, is considered. The inclusion properties to be identified are the material properties - elastic moduli and mass density, and the shape (the interface curve that separates the inclusion from the rest of the domain). This work extends [1] in two ways: firstly, the governing equations are the equations of elasticity rather than the scalar wave equation, and secondly, the scatterer is an inclusion and not a hole. In the proposed method, the identification is performed using full waveform inversion. Sensors measure the waves generated by known tractions, then, a two-stage iterative process is performed to find the inclusion's properties that best matches the measurements. The adjoint method is used for its efficient calculation of the gradient of the misfit function. The unknown interface is defined by a parametric representation, that is completely general and does not make any preliminary assumptions about the scatterer geometry. The performance of the proposed method is demonstrated via numerical examples, starting with material property identification alone, followed by shape identification alone, ending with simultaneous identification of both material properties and shape. All numerical experiments were performed on inclusions with varying contrasts: inclusions with material properties close to those of the background, and inclusions with significantly different material properties from the background. It is concluded that the contrast of the inclusion has a significant impact on the success of the identification.

REFERENCES

[1] Sayag, A., and Givoli, D. Shape identification of scatterers using a time-dependent adjoint method. Computer Methods in Applied Mechanics and Engineering 394 (2022): 114923.

Hourly Scale Model of Wind Magnitude and Direction Based on Stochastic Differential Equations

Maayan Shimoni
Advisor- Dr. Anna Clarke
Faculty of Aerospace Engineering
Technion - Israel Institute of Technology

April 7, 2024

Abstract

Wind modelling is relevant to many fields, including aircraft performances, wind energy estimation, fatigue loads analysis, and many others. Wind behavior is stochastic, influenced by factors such as season, geographical and meteorological conditions, and topography. Hence, wind has a wide range of behaviors and it can be analyzed and modelled in many ways. The main goal of this work is building a long-scale model that will provide one value per hour for both magnitude and direction of wind velocity. These values are essentially the mean values of the hourly period.
Both magnitude and direction are modelled using a known method based on stochastic differential equations (SDEs). This method enables to produce a stochastic process with any desired probability distribution, using a memoryless transformation. The processes produced by the method have an exponentially decaying autocorrelation. Although this method is used in literature exclusively for analyzing the wind velocity's magnitude, in this work it is adopted to the direction analysis as well.
Wind speed is usually modelled by non-Gaussian distributions. For instance, the Weibull distribution is appropriate for modelling wind speeds, since it has shown a good fit to observed wind speed data in many locations around the world in a time scale of hours. Therefore, this distribution is chosen for modelling the magnitude of the wind velocity.
The direction of the wind is modelled using a mixture of Von Mises distributions, which is the circular analogue of the normal distribution. This is chosen due to the flexibility of the model that can represent many different wind regimes. Real world data from a meteorological station in New Zealand, Wellington, taken from the New Zealand national climate database, is analyzed, and the model is tuned to this data. A comparison between the statistical properties of the data and the stochastic process generated by the model is also provided.

Optimization of Aircraft Stringer Reinforcements

S.E. Katzeff[a]

[a]*Principal Stress Engineer. Israel Aerospace Industries: Upgrades and Conversion Division, Lod, Israel.*

Abstract

Aircraft stringers are designed to 'carry' axial force due to fuselage bending. Stringer design is focussed on optimizing its cross-section area to prevent crippling, as well as its inertia to prevent buckling. A symmetrical cross-section is therefore not essential, as the stringer should not 'see' any applied bending. However as shown in Figure 1, due to the installation of reinforcements, which usually embodies attaching more area to the stringer over a relatively long portion, a shift in the neutral axis of the built-up section occurs. This inevitable change in neutral axis results in an offset moment applied to the stringer-skin section. As the stringer is attached (via fasteners or bonded) to skin at its outer flange, it is prevented from displacing in the transverse direction. Hence, the stringer is considered stabilized, and any offset moment is reacted along its principal horizontal axis. Many stringer profiles are un-symmetrical, and hence employ a more complex manner of reacting moments. Peery, in chapter 7 of his Aircraft Structures, outlines the reduction in the complex bending formula to obtain a moment, applied transverse to the principal horizontal axis, that is required to stabilize such un-symmetrical cross sections. If the principal horizontal axis offset bending moment can be considered parasitic in nature, then the further applied moment in the transverse direction is even more so. Assuming a fastened section, this stabilizing transverse moment must be applied by the fasteners connecting the stringer to the skin over a short distance, and results in a drastic increase in shear force at the first and last 3 fasteners, as shown in Figure 2.

Choosing a concept that minimizes this resultant shear force is therefore a further consideration that should be evaluated when designing stringer reinforcements. Ignoring both resulting parasitic bending moments can have dire consequences for both static margins and fatigue life of the attachment to the skin. In the proposed investigation, the shear force at the stringer-skin fasteners due to reinforcement of un-symmetrical stringer sections is determined using analytical methods, and is validated by FEA. The particular case of the stringer stabilized at its outer flange only is considered, and the resulting transverse bending moment and internal stress distribution is determined by minimizing the internal strain energy. The proposed investigation concludes that the best/worst solutions for minimizing the impact of the offset moment in an un-symmetric stringer is to reinforce in such a way that the neutral axis is moved along its maximum/minimum principal axis respectively. In addition, it is recommended that the first/last 3 fasteners connecting the stringer to the skin at the reinforcement should always be enlarged (or in the case of rivets, changed to Hi-Loks) to account for the sudden applied transverse force. As a final recommendation, F&DT inspections should also be performed at reinforcement locations at the first/last 3 stringer-skin fasteners.

Keywords: Complex bending, Stringer reinforcement, Offset moment, Stabilized stringer

Email address: skatzeff@iai.co.il (S.E. Katzeff)

Preprint submitted to IACAS Conference 2024 *May 1, 2024*

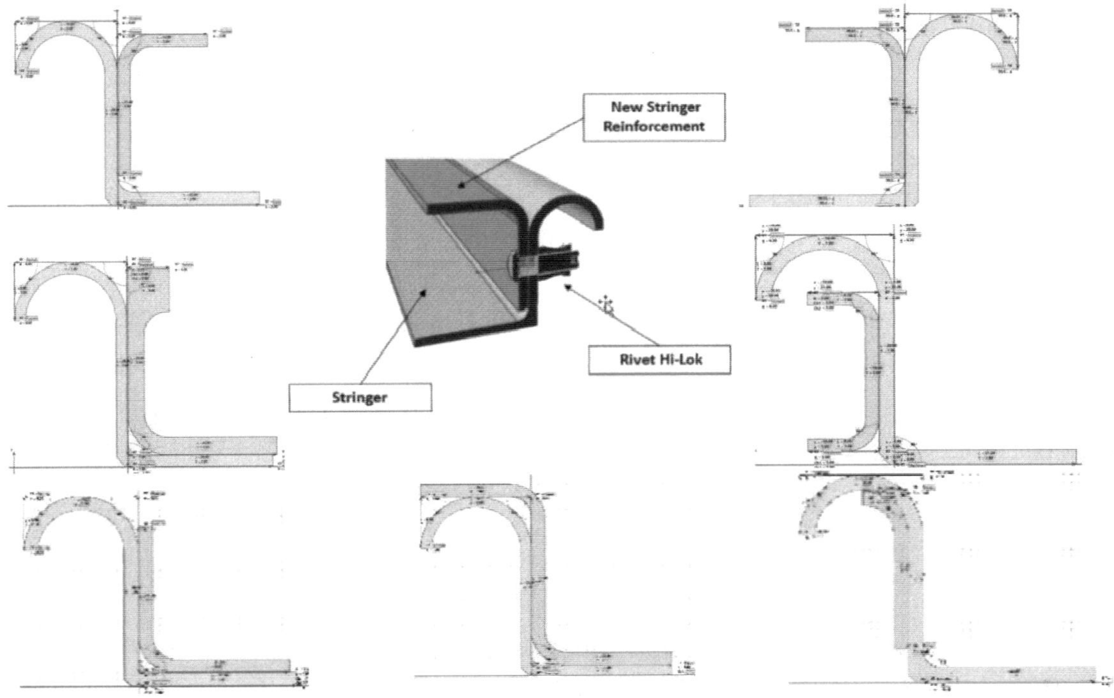

Figure 1: Typical options of stringer reinforcement

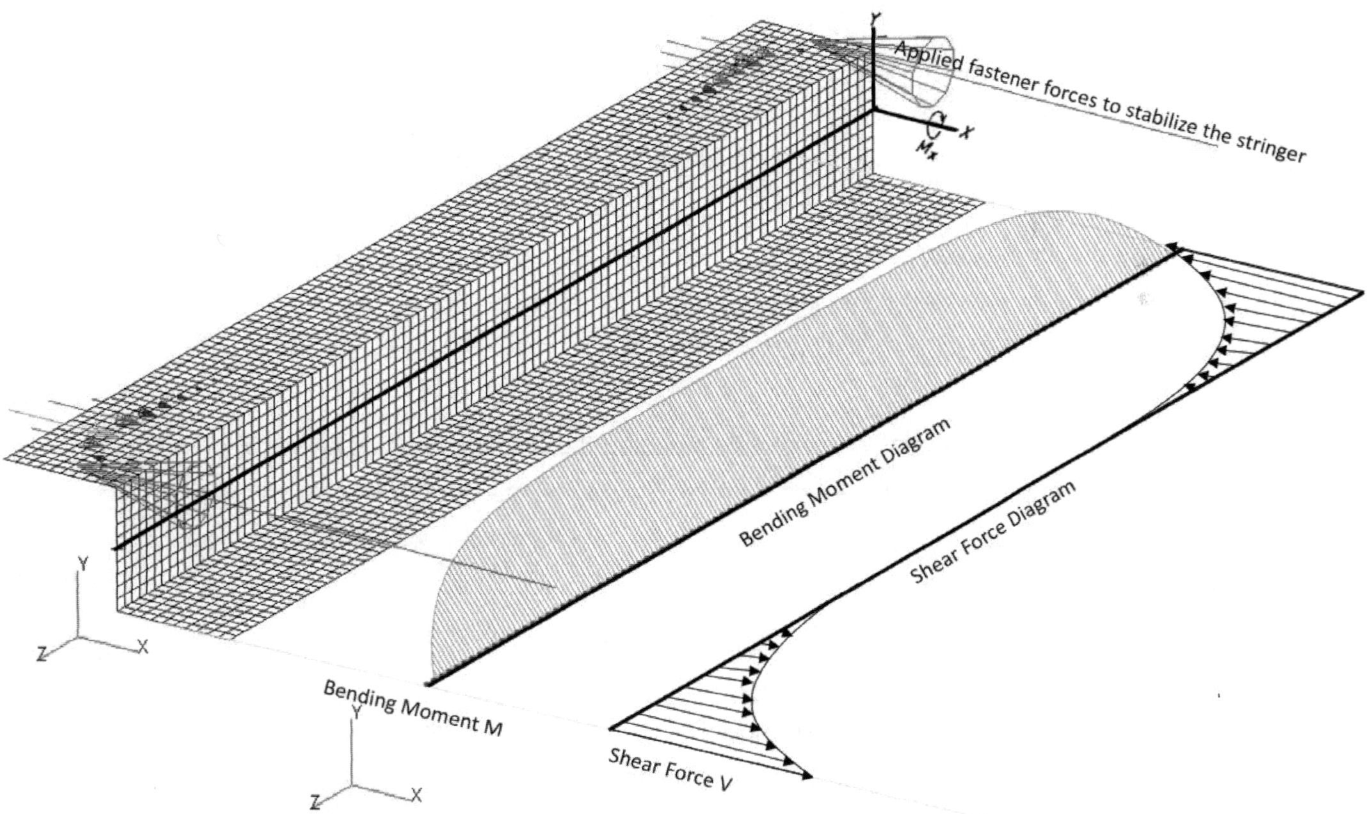

- The fasteners at the skin apply a transverse force to stabilize the stringer;
- This forces are largest at the first and last 3 fasteners;
- The transverse fastener forces should be considered when deciding how best to reinforce the stringer;
- The transverse fastener forces should also be considered in the static / F&DT justification of the reinforcement.

Figure 2: Free-body, shear and bending moment diagram of reinforced stringer

DriveSat Student Project

Authors: Noa Offri, Nitzan Schwartz

Supervised by: Jacob Herscovitz

The DriveSat project is an ongoing undergraduate final project aimed at developing a 2U CubeSat for imaging missions in low Earth orbit (LEO). Central to its design is the innovative use of HDD reaction wheels (HDD-RW), proposed by ASRI researchers as a cost-effective alternative to traditional reaction wheels. Recently, a camera for LEO imaging was incorporated as a secondary payload. DriveSat's primary objectives include demonstrating the viability of HDDs as reaction wheels in space environments and conducting imaging tasks to test attitude control using HDD-RW.

To achieve these goals, the missions and orbits team devised scenarios categorized into four main modes: Detumbling, Communication, Imaging, and Cruise. Utilizing tools like STK facilitated data analysis tailored to each team's requirements. Accessing data from the ASRI Ground Station (GS) revealed intricate patterns crucial for verifying telemetry transmission times, particularly beneficial for the communication team.

The 1U payload bay was ingeniously designed to accommodate three HDD reaction wheels, the camera, and corresponding adapters for integration into the satellite frame. These adapters ensure payload bay resilience during launch conditions. Selecting a suitable camera involved considerations such as resolution, power efficiency, and longevity, leading to the choice of Gomspace's C1U with a 70mm lens.

The avionics team's primary focus lay in simulating the power system, furnishing vital insights into battery status, power consumption, and generation. Additionally, a data downloading simulation based on access times and data collection rates was developed, enhancing mission analysis capabilities.

Attaining satellite attitude control across various operating modes—detumbling, Sun-cruise, night-cruise, and imaging—was paramount. A desaturation controller leveraging magnetorquers to desaturate HDD reaction wheel actuators was integrated. Detailed and simplified block diagrams of the satellite's software were crafted, alongside a software requirements document for future implementation on the FlatSat, incorporating an anti-deadlocking algorithm for remote override.

Effective communication with the satellite during its mission relies heavily on the ground station. Consequently, comprehensive requirements and operations documents were formulated to ensure smooth ground station operation. Integrating the efforts of all groups and simulating the satellite holistically posed a significant challenge for the integration team. This involved incorporating data from the orbit team's propagator, ADCS architecture, power system, ground station communications, camera, and software, culminating in the development of a new GUI for simulation interfacing.

In recapitulation, the project yielded a multitude of noteworthy achievements. Foremost among them was the successful demonstration of the functioning Reaction Wheel (RW) within a vacuum chamber, a milestone underscoring its viability in space environments. Equally significant was the attainment of angular velocity decay to zero through adept utilization of control algorithms, a feat crucial for precise maneuvering in orbit. Additionally, the project introduced strides in the realm of structural design, marked by the meticulous planning analysis and printing of a structural model. Moreover, the comprehensive system simulation was completed fully alongside the preliminary design of all satellite subsystems, embodying a foundational step towards DriveSat's advancement. Undoubtedly, the project's progression underscores the indispensable necessity for interdisciplinary collaboration to propel DriveSat towards its envisioned goals.

Analytical Dynamics Within Rotating Frames: Addressing the Keplerian Ballistic Problem

Iris Kanter* and Vladimir Martinusi†

Department of Aerospace Engineering
Technion – Israel Institute of Technology

Abstract

This study delves deeply into the intricate analytical modeling of gravitational motion within a rotating reference frame, with a specific focus on addressing the terrestrial ballistic problem governed by Keplerian dynamics. The rotational motion of the Earth adds complexity to this gravitational challenge, rendering the attainment of an analytical solution unfeasible through conventional means. In response, this research pioneers the application of numerical methodologies to tackle this formidable task.

At the heart of this endeavor lies the development of a novel fixed point iteration algorithm meticulously crafted to identify the precise initial conditions required for launching a ballistic missile towards a predetermined distant target, while simultaneously adhering to stringent energy constraints. Through rigorous testing and refinement, the algorithm demonstrates remarkable efficacy, showcasing rapid convergence towards the exact initial inertial velocity necessary for accurate targeting. Remarkably, the achieved accuracy is within a mere tens of meters relative to the specified terrestrial target, highlighting the algorithm's exceptional precision and reliability.

Furthermore, this study serves as a cornerstone for the future advancement of ballistic missile interception strategies. By laying a robust foundation for the development of advanced algorithms, it sets the stage for the creation of optimized interception techniques capable of effectively neutralizing potential threats. By harnessing the power of numerical computation and innovative algorithmic design, this research not only pushes the boundaries of gravitational modeling but also holds immense promise for enhancing global security and defense capabilities.

*Undergraduate Student. Email: iriskanter@campus@technion.ac.il

†Senior Visiting Lecturer. Email: vmartinusi@technion.ac.il

Employing Intermediate Points for Optimal Low-Thrust Docking with a Constrained Approach Direction

Or Nahum* and Vitaly Shaferman†
Technion, Israel Institute of Technology, Haifa, Israel, 3200003

In this paper, a new optimal-control-based guidance law for a perfect rendezvous from a predetermined terminal approach direction, i.e., docking, is derived. The guidance law softly constrains the chaser to pass through intermediate points located in the desired approach direction, in addition to the rendezvous soft constraint at the end of the scenario and the minimization of the control effort. As a result, the chaser tends to keep its motion in the desired approach direction until reaching the target. The guidance law is derived from a linear quadratic optimal control problem, and the Clohessy-Wiltshire (CW) equations are used to describe the dynamic model. An analytical, open-loop solution is derived, and a closed-loop guidance law is presented as well. The designer determines the intermediate points' locations, while the speeds at these points are obtained by a secondary optimization process whose performance index is also determined by the designer. The paper presents two secondary performance indexes: The original problem's cost function and the minimum deviation indexes. Simulations demonstrate the proposed guidance law's performance, showing small miss distances, velocity misses, and angle misses. It is concluded that the new proposed guidance law brings the chaser to a perfect rendezvous with the target in the desired approach direction. Moreover, the approach trajectory can be shaped via different secondary optimization criteria. In addition, the simulation study shows that the additional terminal approach direction constraint added in this paper only slightly increased the overall control effort compared to the unconstrained classical optimal perfect rendezvous guidance law.

*Graduate Student, Faculty of Aerospace Engineering, Technion, or.nachum@campus.technion.ac.il.
†Associate Professor, Faculty of Aerospace Engineering, Technion, vitalysh@technion.ac.il.

Long-Term Mitigation of Earth Oblateness Effects on Large Aperture Satellite Constellations

Ilay Lazarovich* and Vladimir Martinusi†

Department of Aerospace Engineering
Technion – Israel Institute of Technology

Abstract

Large Aperture Satellite Formations (LASF) constitute a critical component in the realm of space-based observational and communicational missions, enabling a spectrum of applications from remote sensing to global communication networks. However, maintaining the precise relative distances among satellites within these formations poses significant challenges, particularly over extended periods, primarily due to gravitational perturbations. Among these perturbations, Earth's oblateness, represented by the J2 gravitational harmonic, emerges as a principal source of disturbance.

This study proposes a novel methodology harnessing Hamiltonian normalization and algebraic Hamiltonian reduction techniques to mitigate the effects of Earth's oblateness on satellite formations. By employing analytical formation initialization, the proposed approach offers a robust solution capable of preserving long-term stability without perceptible drift for durations spanning up to a year, regardless of the mutual distances between the satellites. This advancement holds immense promise for enhancing the operational longevity and reliability of LASF missions, as it addresses a fundamental challenge in maintaining formation integrity.

Furthermore, this strategy lays the groundwork for the development of highly efficient control algorithms aimed at minimizing the fuel consumption required to uphold the desired distance thresholds between satellites within the formation. By leveraging analytical techniques to counteract gravitational perturbations, this approach not only enhances the overall resilience of satellite formations but also facilitates the optimization of resource utilization, thereby potentially reducing mission costs and extending operational lifetimes. Consequently, this research contributes significantly to the advancement of space-based technologies, opening avenues for the realization of more sustainable and economically viable satellite missions in the future.

---

*Email: ilaylazarovich@gmail.com

†Senior Visiting Lecturer. Email: vmartinusi@technion.ac.il

AI-Empowered Resilient Quadrotor Navigation

Itzik Klein[*]
The Autonomous Navigation and Sensor Fusion Lab, Hatter Department of Marine Technologies, University of Haifa, Israel

In the last decade, quadrotor usage for civilian and military applications has significantly increased in many applications. Those include construction, deliveries, surveillance, industry, marine science, mapping, emergency response, and law enforcement [1]. To accomplish the challenging tasks assigned to them, quadrotors require an accurate navigation system. While outdoors, the fusion of a global navigation satellite systems (GNSS) receiver with an inertial navigation system (INS) is one of the commonly used approaches for quadrotor navigation. A model-based extended Kalman filter framework is commonly used for this type of fusion. This talk presents two challenges in the field and discusses ways to overcome them.

1) **QDRNet**: For various reasons, such as satellite signal blocking, the quadrotor's navigation solution depends only on the inertial navigation system solution. Consequently, the navigation solution drifts in time due to errors and noises in the inertial sensor readings. To mitigate the inertial drift we present QDRNet [2] hybrid framework for quadrotor dead reckoning to estimate the quadrotor's three-dimensional position vector at any user-defined time rate. As a hybrid approach, QuadNet uses both neural networks and model-based equations during its operation. QuadNet requires only the inertial sensor readings to provide the position vector.

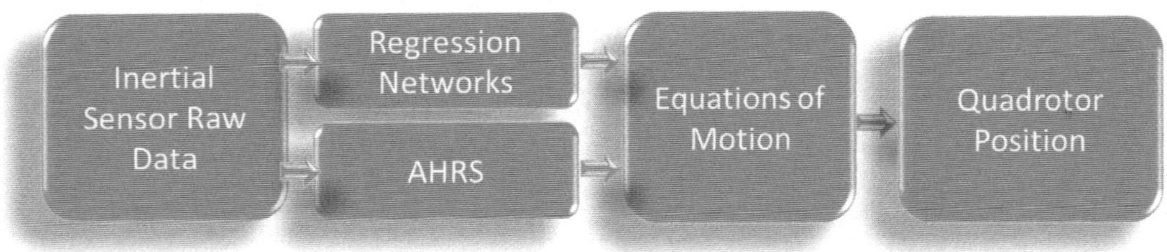

Figure 1. QDRNet hybrid structure for estimating the quadrotor position in situations of pure inertial navigation.

2) **ProNet**: One of the important parameters of the filter is the process noise covariance. It is responsible for the real-time solution accuracy, as it considers both vehicle dynamics uncertainty and the inertial sensors quality [3]. Typically, the process noise covariance is assumed constant. Yet, due to vehicle dynamics and sensor measurement variations throughout the trajectory, the process noise covariance is subject to change. We offer a hybrid model and learning-based adaptive navigation Kalman filter [4,5]. We rely on the model-based Kalman filter and design a deep learning model to adjust the momentary system noise covariance matrix, based only on the inertial sensor readings. Once the process noise covariance is learned, it is plugged into the well-established model-based extended Kalman filter.

[*] Email: kitzik@univ.haifa.ac.il

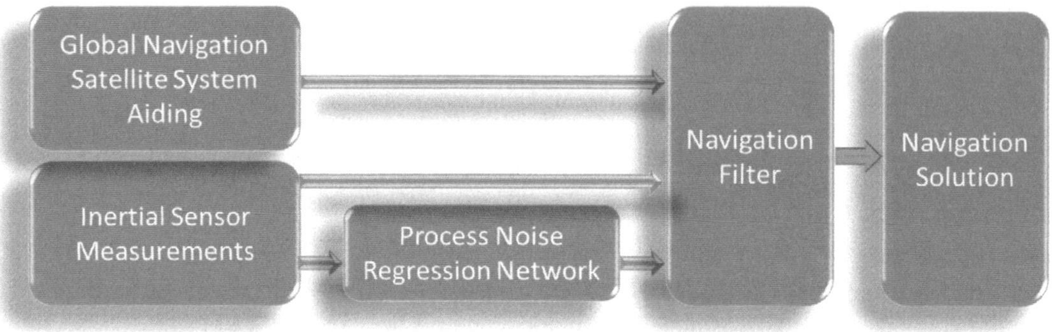

Figure 2. ProNet hybrid adaptive navigation filter block diagram.

.

References

[1] Idrissi, M., Salami, M., and Annaz, F., "A Review of Quadrotor Unmanned Aerial Vehicles: Applications, Architectural Design and Control Algorithms," Journal of Intelligent and Robotic Systems, vol. 104, no. 2, pp. 1-33, 2022.

[2] Shurin, A. and Klein, I., "QDR: A Quadrotor Dead Reckoning Framework," IEEE Access, vol. 8, pp. 204433–204440, 2020.

[3] Zhang L., Sidoti D., Bienkowski A., Pattipati K. R., Bar-Shalom Y., and Kleinman D. L., "On the identification of noise covariances and adaptive Kalman filtering: A new look at a 50 year-old problem," IEEE Access, vol. 8, pp. 59362–59388, 2020.

[4] Klein I., "Data-Driven Meets Navigation: Concepts, Models, and Experimental Validation," Inertial Sensors and Systems, Braunschweig, Germany, 2022.

[5] Solodar D. and Klein I., "VIO-DualProNet: Visual-Inertial Odometry with Learning Based Process Noise Covariance", Engineering Applications of Artificial Intelligence, vol. 133, pp. 108466, 2024.

Anomaly Detection in Structural Vibration-Tests through Machine-Learning Processes on Accelerometer Measurements

Shiran Tsolker, Shahar Yehezkel, Ido Hauzer, Ariel Drachinsky *
RAFAEL Advanced Defense Systems, Haifa, 31021, Israel

This study proposes a machine learned based method to automatically identify anomalies during mechanical vibration tests.

Vibration testing is a vital component in the development and certification process of aerospace structures. These physical tests are performed on a manufactured prototype during the development phase and are hence a significant project-milestone risk in case of unforeseen structural failure.

A typical vibration test is conducted by mounting the structure on an electrodynamic shaker and exciting it according to a specified vibration profile, while measuring acceleration responses and sometimes strains in designated locations.

When failure occurs in such tests the damage often begins as a local defect that is hard to monitor or predict yet may lead to changes in the system's response. If not detected in time, the damage often propagates to multiple locations, significantly increasing the extent and cost of the damage complicating the failure study.

To address this detection issue, vibration-tests may be monitored in real-time with the aim of detecting changes in the structural response, thus indicating a possible deterioration of the structural integrity prior to failure propagation [1].

This work compares two machine-learning-based algorithms for structural anomaly detection during vibration tests, reliant on the assumption that at the beginning of the test the structure is uncompromised.

Both algorithms start by extracting scalar features from windows of time-history accelerometers' measurements. The features include time-dependent statistical and spectral characteristics of the measured signals. The scalar values of all features comprise the state-vector of the system.

The 1st is a distribution-based algorithm, which learns the distribution of the state-vectors as a multi-dimensional gaussian distribution during the training phase and then ranks each of the following time-windows as the vibration-test progresses, where the rank of each window represents the Mahalanobis distance of the current window from the multidimensional gaussian distribution [2]. Once a state-vector is found to be outside a threshold distance from the distribution means, it is marked as anomalous.

The 2nd is a clustering-based algorithm,[3]. It creates an embeddings vector through Principal-Component-Analysis (PCA) of the state vectors and performs a clustering operation on the reduced embeddings-space, by a density-based algorithm that assigns each time-window to a cluster. With the appearance of a new cluster, it is considered anomalous.

* Mechanical engineer, shirants@rafael.co.il

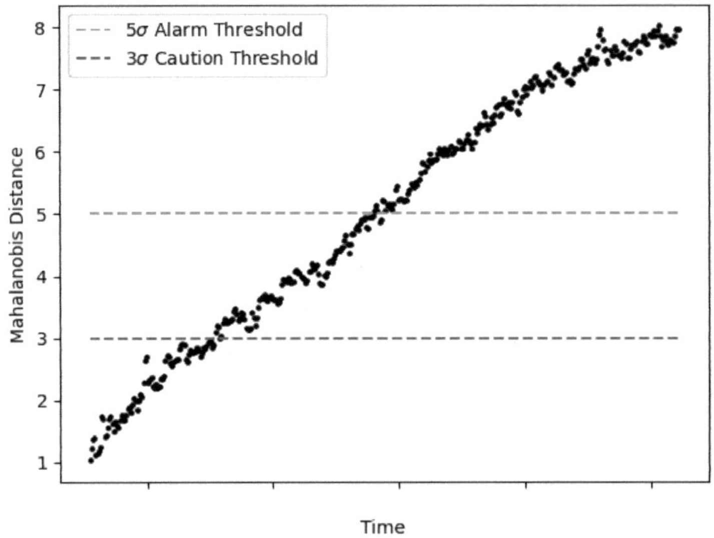

Figure 1. Anomaly identification using density based algorithem

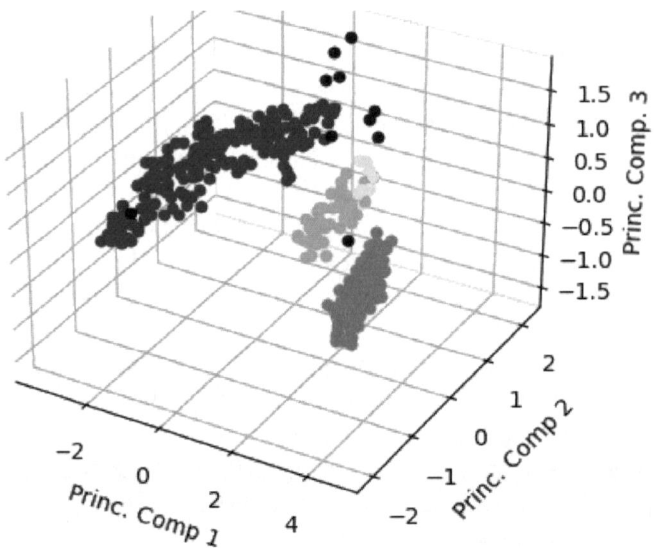

Figure 2. Results of a clustering operation performed on PCA-reduced data.

Preliminary Results & Future Work

Both algorithms have been deployed on data acquired from vibration-tests on operational structures, few of which resulted in structural failure. The density-based algorithm exhibited near 100% success in classifying the tests where anomaly occurred, with no false positives, while the clustering-based algorithm classified a significant portion of the structurally-sound tests as ones where an anomaly occurred.

Though some work still remains to optimize the correct and early detection of structural anomalies and minimize false-alarms, the algorithms are to be embedded in real time lab tests monitoring to ensure that mechanical failures in vibration-tests are detected early enough to minimize damage and ease the failure studies when needed.

References

[1] - Toh G, Park J. Review of Vibration-Based Structural Health Monitoring Using Deep Learning. Applied Sciences. 2020; 10(5):1680. https://doi.org/10.3390/app10051680

[2] - Yeager M, Gregory B, Key C, Todd M. On using robust Mahalanobis distance estimations for feature discrimination in a damage detection scenario. Structural Health Monitoring. 2019;18(1):245-253. doi:10.1177/1475921717748878

[3] - Zhang Y, Li Y, Kong L, Niu Q, Bai Y. Improved DBSCAN Spindle Bearing Condition Monitoring Method Based on Kurtosis and Sample Entropy. Machines. 2022; 10(5):363. https://doi.org/10.3390/machines10050363

הרצאה מפגש IACAS 2024 טכניון

GAIA – Generative AI for Aerospace

מרצים: ארז שרון B.Sc. – התעשייה האווירית לישראל
אביב פריד B.Sc. – התעשייה האווירית לישראל

תקציר

בינה מלאכותית גנרטיבית (Generative AI) - הינו תת תחום של בינה מלאכותית המתפתח במהירות, מחולל מהפכה של ממש באופן בו אנו תופסים ויוצרים אינטראקציה עם טכנולוגיה. התחום מאפשר הגדלת היצירתיות האנושית ויצירת תוכן במהירות ובאיכות מדהימה.
שילוב Generative AI בעבודה השוטפת שלנו יכול להיות **"מכפיל כוח"** משמעותי בתחומים שונים כמו פיתוח, אנליזה וארכיטקטורת תוכנה, סיכומים, כתיבת מסמכים טכנולוגיים, כתיבת קוד ואוטומציה, עזרה בפתרון בעיות טכניות, אופטימיזציה באלגוריתמים, זיהוי שגיאות ועוד.

GAIA, ראשי תיבות של Generative AI for Aerospace, מציגה קפיצת מדרגה באינטגרציה של AI בתעשיית התעופה והחלל, המותאמת במיוחד לעמוד בדרישות הקפדניות והייחודיות של תחום זה. GAIA רותמת את הכוח של מודלי שפה גדולים ומתקדמים (LLMs) המותאמים לידע וסביבות נתונים ספציפיים של הנדסת תעופה וחלל ובנוסף לכך מתעלה על מגבלות כלי בינה מלאכותית מסחריים קיימים על ידי הצעת פתרון מאובטח, מקומי ומותאם המאפשר עבודה עם מידע רגיש או מסווג.

שילוב AI חדשני ומותאם, הופך את GAIA לכלי נדרש בארגוני תעופה וחלל. GAIA מספקת תשובות ומפיקה תוכן ממסמכים שסופקו על ידי המשתמש ומשלבת ידע הנדסי ייחודי בהתאם לתהליכי העבודה, מה שמבטיח רלוונטיות ודיוק גבוה.
אחד היתרונות המשמעותיים ביותר של GAIA טמון בהשתלבותה העמוקה בתהליכי עבודה פנימיים, בייעול זרימת העבודה, צמצום זמן ומשאבים המושקעים במחקר, עבודה הנדסית, תיעוד ויצירת קוד. פתרונות ניהול הידע המותאמים של GAIA מאפשרים גישה מהירה למידע קריטי, איכות מסמכים משופרת ופיתוח תוכנה מהיר יותר, ומתורגמים ישירות לחיסכון בעלויות ופרודוקטיביות בפרויקטים שונים בתחום התעופה והחלל.

בהרצאתנו, תחילה נסקור מושגי יסוד מעולם הבינה המלאכותית ברמת על ובפרט את מיקומה של ה-Gen AI ומודלי שפה גדולים LLM, נציג תהליכי עבודה בהם שילבנו את GAIA ונציג איך יישום המערכת מביא לשיפור פרודוקטיביות וצמצום עלויות בארגון כמו תע"א.

לסיכום, GAIA עומדת בחזית שילוב הבינה המלאכותית בתעופה וחלל, ומציעה פתרון מותאם תוך מענה לדרישות הייחודיות של התעשייה וגם סוללת את הדרך להתקדמות משמעותית בניהול הידע הארגוני.
עם GAIA, תע"א יכולה להגיע לרמות חדשות של יעילות, יצירתיות וחדשנות בתהליכי העבודה הפנימיים, ולגבש יתרון תחרותי בתחום העיסוק העיקרי שלה.

Ben Gurion International Airport, 70100 Israel www.iai.co.il

2024 IACAS Technion lecture
Smart Maintenance
Lecturer: Keren Liadski - IAI

ABSTRECT

Airlines love their planes in the air. They are prepared for the planes to be grounded for two reasons only: when they land, unload and load people, or when they go in for periodic maintenance, which is known and fixed.

The field of maintenance management has undergone many transformations in recent years, when, following the advancement of technology, it is possible to take a proactive approach and perform real-time health monitoring of systems - using advanced tools, such as smart and precise sensors, high-speed communication, powerful computing based on cloud systems, ML and advanced analytical tools.

A preventive maintenance method - which is usually determined based on manufacturer recommendations, analyzes and best practice, usually does not consider the condition of the items, the nature of its operation and working conditions. Now it is possible to harness advanced technologies to predict and prevent potential maintenance problems in systems, before they escalate into expensive and dangerous problems while using AI tools in the spirit of Industry 4.0.

According to this maintenance method, we can notice when there is a decrease in the performance of a certain system and predict faults before they occur. This proactive approach means that we can schedule maintenance during optimal operational windows, by performing focused and spot-on treatment at an optimal time, and prevent the formation of the fault and its development into damage that requires corrective maintenance, as well as save expensive downtime and improve the availability and safety of the system.

This strategic approach to maintenance not only saves us money in the long run, but also optimizes the allocation of resources, reduces waste and maximizes the lifespan of the systems.

In my lecture, we will review our predictive maintenance mechanism and data-driven insights supported by domain experts that allow us to anticipate maintenance needs accurately. By harnessing data-driven insights and advanced technologies, we are changing the way we approach maintenance in all disciplines & a POC in the field of civil aviation will be presented.

Ben Gurion International Airport, 70100 Israel www.iai.co.il

Asynchronous sampled-data synchronization with small communications delays

Gal Barkai, Leonid Mirkin, and Daniel Zelazo

***Index Terms*—Sampled-data systems, network control systems, time delays.**

I. Abstract

The study of multi-agent systems (MAS) has been a central tenant in a multitude of scientific and engineering disciplines over the past decades, with applications including biology, optimization, and robotics. Perhaps the most widely known control problem in MAS is the consensus [1], [2] or synchronization [3], [4] problem, in which the agents must *agree* on some common steady-state trajectory. The crux of this problem is that the agents must come to agreement distributively with limited communication, limitations which may be divided into two categories: spatial and temporal.

Spatially constrained communication reflects the requirement that each agent must generate its control law based on partial information. Each agent is only privy to information from a subset of the group, its *neighbours* [5], [6]. These neighbourhoods can be time-varying and induce a *graph* structure on the overall system, which has been widely exploited to design control laws [7]. While spatial constraints restrict the *identity* of interacting agents, temporal constraints restrict the communication *time*. These constraints encompass both sampled communication and various forms of time-delays, both staples of networked systems.

In practice systems must cope with both types of constraints. Moreover, common assumptions such as continuous communication, static graphs, or periodic synchronous sampling [8]–[10] might not be realistic. This cumulative burden makes even consensus of simple integrators challenging [11], often leading to conservative, robustness-based, approaches [12], [13] even without delays. To the best of our knowledge, achieving synchronization under asynchronous communication and heterogeneous transmission delays is still an open problem.

We propose a sampled-data control protocol that guarantees asymptotic synchronization for switching graphs, asynchronous sampling, and time-varying, heterogeneous transmission delays. Other than standard stabilizability assumption we require only persistent connectivity of the graphs induced by the sampling sequence, which can be asynchronous and unknown. As for the delays, we simply assume that the incoming information is time stamped and the delays are small. More precisely,

$\mathcal{A}_1$: let $\tau_{ij}[k]$ denote the transmission delay from agent j to agent i at time instance $t = s_k$, then we assume that

$$s_k + \tau_{ij}[k] < s_{k+1}, \quad \forall i, j \in \mathbb{N}_\nu,\ k \in \mathbb{Z}_+.$$

The control structure is an extension of the emulation-based hybrid controller proposed in [14], with an added internal predictor. Synchronization is made possible by leveraging the hybrid nature of the original controller, which couples the agents only when information arrives. As a result the delays are akin to a known mismatch between the real state and the delay-free one, which is then compensated by a predictor.

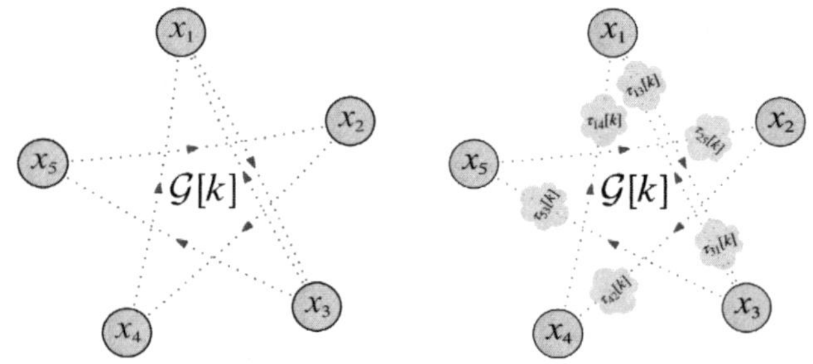

(a) Standard. (b) With delays.

Fig. 1. Sampled-data multi-agent control architectures.

Supported by the Israel Science Foundation (grants no. 3177/21 and 2285/20). GB and LM are with the Faculty of Mechanical Eng. and DŽ is with the Faculty of Aerospace Eng., all at the Technion—IIT, Haifa 3200003, Israel. E-mails: *galbarkai@campus.technion.ac.il*; {*mirkin, dzelazo*} *@technion.ac.il*.

References

[1] R. Olfati-Saber, A. Fax, and R. M. Murray, "Consensus and cooperation in networked multi-agent systems," *Proc. IEEE*, vol. 95, no. 1, pp. 215–233, 2007.

[2] W. Ren and R. W. Beard, *Distributed Consensus in Multi-vehicle Cooperative Control: Theory and Applications*. London: Springer-Verlag, 2008.

[3] L. Scardovi and R. Sepulchre, "Synchronization in networks of identical linear systems," *Automatica*, vol. 45, no. 11, pp. 2557–2562, 2009.

[4] P. Wieland, R. Sepulchre, and F. Allgöwer, "An internal model principle is necessary and sufficient for linear output synchronization," *Automatica*, vol. 47, no. 5, pp. 1068–1074, 2011.
[5] A. Jadbabaie, J. Lin, and A. S. Morse, "Coordination of groups of mobile autonomous agents using nearest neighbor rules," *IEEE Trans. Automat. Control*, vol. 48, no. 6, pp. 988–1001, 2003.
[6] J. A. Fax and R. M. Murray, "Information flow and cooperative control of vehicle formations," *IEEE Trans. Automat. Control*, vol. 49, no. 9, pp. 1465–1476, 2004.
[7] M. Mesbahi and M. Egerstedt, *Graph Theoretic Methods in Multiagent Networks*. Princeton: Princeton University Press, 2010.
[8] H. Chu, D. Yue, C. Dou, X. Xie, and L. Chu, "Consensus of multiagent systems with time-varying input delay via truncated predictor feedback," *IEEE Transactions on Systems, Man, and Cybernetics: Systems*, vol. 51, no. 10, pp. 6062–6073, 2021.
[9] W. Jiang, Y. Chen, and T. Charalambous, "Consensus of general linear multi-agent systems with heterogeneous input and communication delays," *IEEE Control Syst. Lett.*, vol. 5, no. 3, pp. 851–856, 2021.
[10] M. Xing, F. Deng, and Z. Hu, "Sampled-data consensus for multiagent systems with time delays and packet losses," *IEEE Transactions on Systems, Man, and Cybernetics: Systems*, vol. 50, no. 1, pp. 203–210, 2020.
[11] X. Ge, Q.-L. Han, D. Ding, X.-M. Zhang, and B. Ning, "A survey on recent advances in distributed sampled-data cooperative control of multi-agent systems," *Neurocomputing*, vol. 275, pp. 1684–1701, 2018.
[12] Y. Wu, H. Su, P. Shi, Z. Shu, and Z.-G. Wu, "Consensus of multiagent systems using aperiodic sampled-data control," *IEEE Transactions on Cybernetics*, vol. 46, no. 9, pp. 2132–2143, 2016.
[13] J. Sun, Z.-S. Wang, and N.-N. Rong, "Sampled-data consensus of multiagent systems with switching jointly connected topologies via time-varying lyapunov function approach," *International Journal of Robust and Nonlinear Control*, vol. 30, no. 14, pp. 5369–5385, 2020.
[14] G. Barkai, L. Mirkin, and D. Zelazo, "An emulation approach to sampled-data synchronization," in *Proc. 62nd IEEE Conf. Decision and Control*, 2023, pp. 6449–6454.

Deep Compression of Neural Networks for Optimal Control Applications

Johannes Diepolder and Joseph Z. Ben-Asher

We present an approach for the sparse training and quantization of neural networks that relies on the deep compression pipeline developed in [1]. This compression pipeline has been demonstrated to effectively reduce the memory footprint and inference cost of neural networks through a combination of sparse training, quantization, and lossless compression. The pipeline in [1] consists of three main stages. In the first stage, the network is trained and pruned. In the second stage, the weights are quantized based on centroids derived from k-means clustering, including retraining. Finally, Huffman Encoding is employed as a lossless compression method to further reduce the memory footprint. Our main contribution is the integration of optimal control methods into this pipeline. Specifically, we utilize a combination of reinforcement-style learning [2] for retraining the centroids in the quantized network and a fine-tuning stage based on a Quadratic Program (QP).

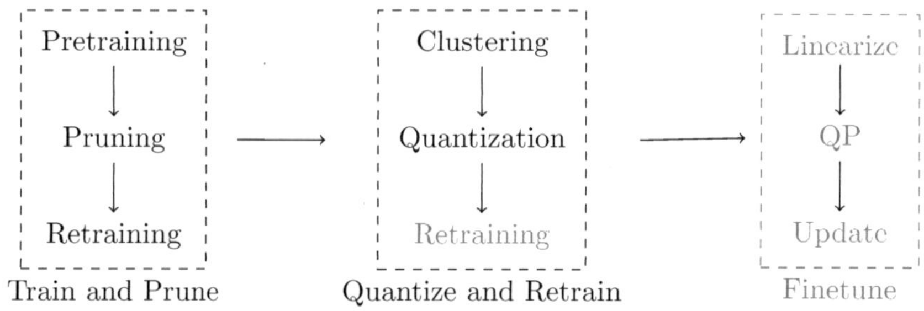

Figure 1: Overview of the adapted compression pipeline for optimal control applications with adapted stages highlighted in blue.

The proposed approach focuses on the supervised training of neural networks using synthetic data generated from parametric optimal control problems. In this context, we essentially leverage the fact that the underlying problem is fundamentally an optimal control problem. On the one hand, the combination of supervised learning on data points derived from optimal solutions with reinforcement learning provides a more direct means of incorporating loss terms regarding the cost and constraints. On the other hand, the errors introduced through parameter pruning and quantization are further compensated for by employing a quadratic update for fine-tuning the results. The approach is illustrated through an aerospace application concerning the midcourse guidance of an aerial vehicle derived from [3]. It is shown that the application of the proposed pipeline exhibits excellent performance while significantly reducing the size of the neural network.

References

[1] Han S, Mao H, Dally W J. Deep Compression: Compressing Deep Neural Networks With Pruning, Trained Quantization And Huffman Coding. ICLR, 2016.

[2] Sutton, Richard S and Barto, Andrew G. *Reinforcement learning: An introduction.* Second Edition, MIT press, 2018.

[3] Grinfeld N, Yaesh I, Ben-Asher J Z. Midcourse Guidance Using Neural Networks, *IFAC-PapersOnLine*, Volume 56, Issue 2, pp 1168-1173, 2023.

Efficient Aerodynamic Database Construction: Leveraging Sparse Sensing and Machine Learning

Shimon Eliahu Julius and Michael Weidenfeld

Aeroballistics Dep., Land Division, Elbit, POB 1044 Ramat Hasharon 4711001, Israel
michael.weidenfeld@elbitsystems.com

Producing an aerodynamic database for configurations operating in a wide spectrum of flow conditions, equipped with multiple aerodynamic control surfaces, presents a substantial challenge. This task traditionally requires extensive computational fluid dynamics (CFD) efforts or a prolonged wind tunnel campaign. This study introduces a method to significantly streamline the process through the application of data science and machine learning (ML) techniques, aiming to decrease the computational cost while preserving the accuracy and integrity of aerodynamic models.

The initial stage involved establishing a benchmark database by executing approximately 5000 CFD computations for a configuration with delta wings and four tail control fins. Importantly, the original CFD computations forming this database were rigorously validated against wind tunnel tests, ensuring their reliability and accuracy.

To facilitate the aerodynamic assessment of slightly modified configurations without resorting to another extensive series of CFD simulations, two methodologies were employed: a machine learning sensitivity analysis approach based on the leave-one-out principle was implemented, whereby each data point in the dataset is systematically removed one at a time, and the model training is performed with the remaining data points. The datapoints that contribute the least to increasing the error of the model prediction are removed at each iteration until the desired error tolerance is reached. This approach was made possible by use of rapid ML techniques, such as XGBoost, which employs gradient boosting to enhance an ensemble of weak decision tree learners into a robust predictive model. Code efficiency was further enhanced through the implementation of multiple distanced data dropouts per iteration and code parallelization. Method accuracy was further improved by modeling the configuration on a 'component' level, thereby aiding the resolution of complex trends in the database. In addition, a data-driven sparse sensing technique was also implemented, utilizing singular value decomposition (SVD) and QR pivoting, to efficiently identify key 'sensor' (datapoint) locations, allowing for a concise yet accurate reconstruction of the aerodynamic database from a minimal set of simulations.

Both the leave-one-out database pruning approach and the sparse sensing method adeptly construct models that regress to the entire 5000-point database, showcasing the potential to greatly reduce the computational burden and time requirements for accurate aerodynamic database production. While both methods significantly reduced the number of datapoints required for a given error tolerance, the leave-one-out method necessitated the use of considerably more computational resources. Meanwhile, the sparse sensing approach can be implemented and promptly executed with just a few command lines in Matlab or Python, for an order of magnitude reduction in dataset size, thus emphasizing the remarkable efficiency of modal decomposition techniques in handling such tasks with minimal resource requirements.

This study demonstrates the effectiveness of ML and data-driven methods for practically rebuilding entire databases for slightly modified configurations with fewer simulations, maintaining accuracy. It highlights the efficiency of data-driven techniques over brute-force approaches, offering insights into optimizing aerodynamic database construction. Both methods prove valuable for their computational economy and precision in modeling aerodynamic changes.

Input-output approach for modeling flow response to actuation in transitional boundary layer

Ofek Frank-Shapir*, Igal Gluzman†

Technion - Israel Institute of Technology, Haifa 3200003, Israel

Predicting the resulting flow field due to actuation is highly important in designing feedback controllers in active flow control applications and studying the flow physics of boundary layers subject to external forcing. We have developed an input-output solver based on linearized Navier-Stokes equations, similarly to the solver of [Jovanovi´c and Bamieh, 2005], to capture structural features and energy pathways of transitional wall-bounded shear flows. This solver has been utilized for three purposes.

First, we used our solver to study the flow physics of energy amplification subject to external periodic forcing for the Blasius boundary layer and plane Poiseuille flow. We find that all norms have a similar general structure to the ones produced in Poiseuille flow, but several key differences have been identified. The flow response to external forcing in any direction in Blasius case resulted in dominating oblique wave structures, whereas in the Poiseuille case, streaky structures were dominant. Generally, all energy norms reach higher peaks for Blasius base flow than for Poiseuille base flow, demonstrating that Blasius flow serves as a stronger amplifier to disturbances than Poiseuille flow.

Second, we have generalized our solver to account for nonlinear interactions by incorporating them within external forcing terms in our input-output solver. This was obtained via different approaches that led to different energy pathways compared to the linear case for both Blasius and Poiseuille base flow cases. We will discuss the different approaches and the resulting differences in the presentation.

Third, we have extended our input-output solver to model the flow response to actuation following [Gluzman and Gayme, 2021] approach, where the actuator geometry was represented as a cluster of single-point source arrays arranged in an actuator's geometrical pattern. We utilize this approach to determine the optimal spacing between a spanwise array of symmetric plasma actuators for maximal amplification of the streamwise velocity component. We propose a novel simplified procedure that exploits the periodic nature of the studied actuator configuration, to reduce the optimization process by several orders of magnitude. Our novel modeling tool can contribute to the understanding of flow control mechanisms and can be utilized in designing and optimizing spanwise-periodic actuation patterns in active flow control applications at a fraction of the time compared to high-fidelity simulations.

References

[Gluzman and Gayme, 2021] Gluzman, I., & Gayme, D. F. (2021). Input-output framework for actuated boundary layers. Phys. Rev. Fluids, 6, 053901.

[Jovanovi´c and Bamieh, 2005] Jovanovi´c, M. R., & Bamieh, B. (2005). Componentwise energy amplification in channelflows. J. Fluid Mech., 534, 145–183.

*Undergraduate Student, Faculty of Aerospace Engineering. E-mail: ofekfs@gmail.com
†Assistant Professor, Faculty of Aerospace Engineering. E-mail: igal.gluzman@technion.ac.il

Effect of Krylov Solver on Dual-Time Stepping Convergence

Ya'akov Mindelis*, Ilya Kislitsin†, Sahar Shpitz‡, and Yuval Levy§
Israeli CFD Center, Caesarea Industrial Park, 3088900, Israel

Time-accurate numerical flow simulations are conducted in support of pitching damping moment (PDM) coefficient predictions. Comparison with experimental data shows that computational tools can be used for that purpose. It is found that high computational efficiency can be obtained by utilizing the Krylov solver and the back-traced line search algorithm for both the steady state solution and for the dual-time stepping iterations. The saving in computational time are very significant as execution time can be reduced by 35-40% in certain cases.

I Introduction

Dynamic stability derivatives (pitch, yaw, and roll), especially damping derivatives on slender projectiles with fins, have a strong influence on the projectile response at extreme maneuver conditions. Theoretical predictions apply only at low angles of attack, and since modern missiles are highly maneuverable, predictions using computational means at large angles of attack are of increasing interest. Such computational means become an essential tool in the design process of slender projectiles. Consequently, efficient capability to numerically calculate these derivatives have become an important feature of any computational fluid dynamics tool. Naturally, validation of the numerical results, in the form of comparison with experimental data, must be completed before the tool is declared operational.

A typical test mechanism utilizes a small-amplitude one-degree-of-freedom cross-flexure balance that is supported by a strut and sting that can be manually adjusted to provide minimum aerodynamic interference and various nominal angles of attack (see Figures 1 and 2, taken from the work by Uselton and Uselton [1], and a schematic of the Basic Finner geometry in Figure 3)

In the experiment, a sinusoidal pitching motion is prescribed such that the angle of attack follows the relation:

$$\alpha(t) = \alpha_0 + A\sin(\omega t) \tag{1}$$

where A is the amplitude and ω is the angular velocity. To computationally obtain the motion driven moment coefficient, time accurate numerical simulations must be conducted. To that end, dual time stepping simulations is employed such that the time-accurate simulation is advanced to the next step only after a certain level of convergence has been obtained. Efficient convergence is therefore essential for the viability of the computational tool.

*Software Developer, jack@iscfdc.com
†Application Engineer, ilya@iscfdc.com
‡Chief Scientist, sahar@iscfdc.com
§CEO, yuval@iscfdc.com

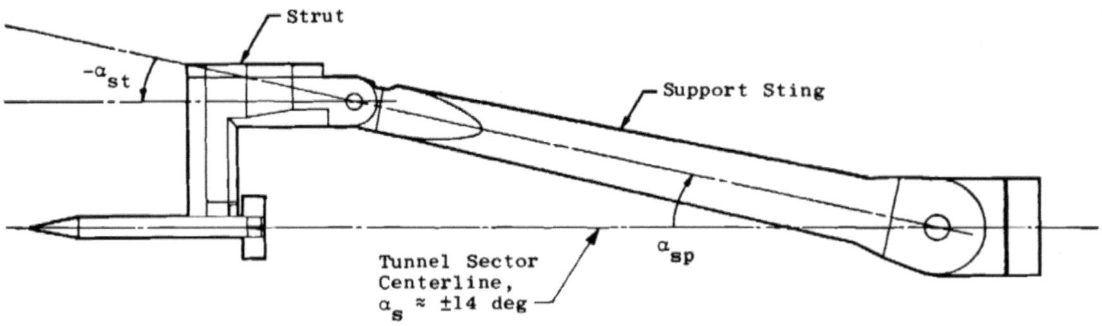

Figure 1. Support sting and strut

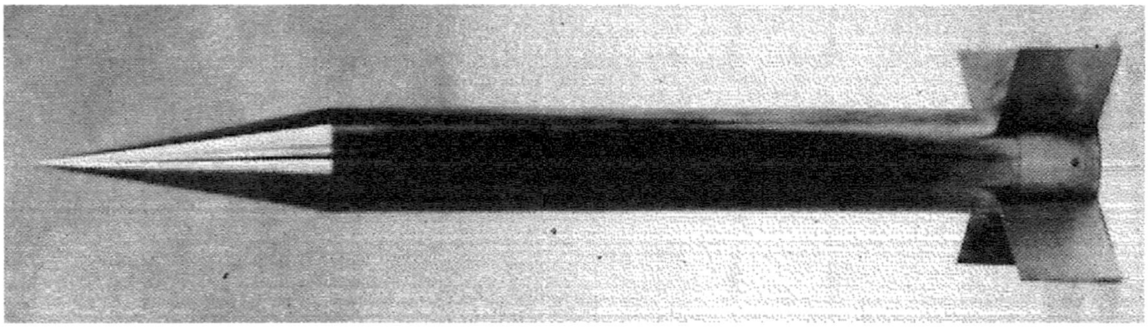

Figure 2. Basic Finner geometry (× configuration)

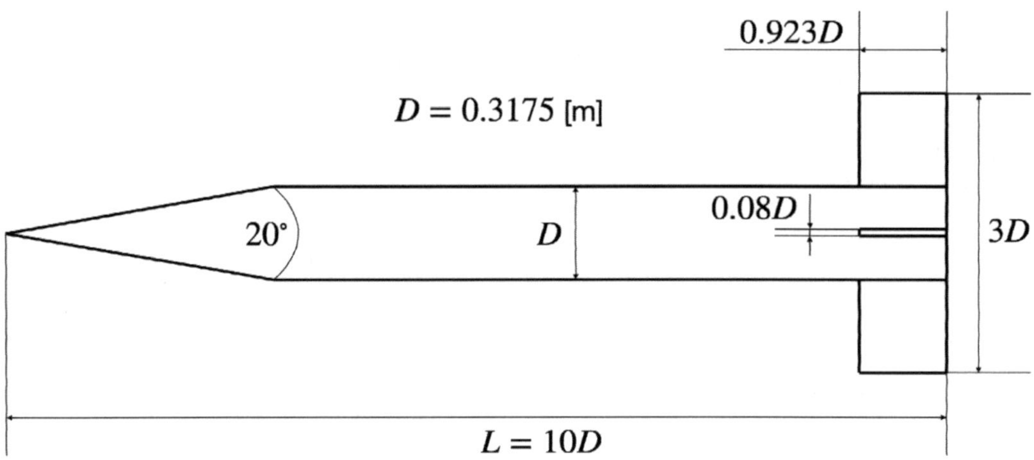

Figure 3. Schematic of the Basic Finner geometry (+ configuration)

II Validation

A series of numerical flow simulations have been conducted to validate the capability to predict the pitching damping moment coefficient. The flow conditions and procedures were chosen to match those of the experiment run by Uselton and Uselton [1]. The free stream Mach number was set to $M_\infty = 1.96$, the free stream pressure was set to $p = 3094.4$ [pa] and the temperature to $T = 173.107$ [K]. These flow conditions resulted in a Reynolds number of $Re_D = 0.086 \times 10^6$. The simple oscillatory pitch motion was also set to match the experiment, namely:

$$\alpha(t) = \alpha_0 + A \sin(\omega t) \tag{2}$$

where the amplitude, A, was set to $A = 1°$ and the angular velocity, ω, was set $\omega = 280$ [rad/s]. The nominal angle of attack was set to $\alpha = 0°$, $\alpha = 10°$, $\alpha = 20°$, and $\alpha = 30°$. Each of these cases, pertaining to a different angle of attack, was simulated in two stages. First, the steady state flow about the Basic Finner at the nominal angle of attack was simulated until convergence. Then the simulation was restarted and the oscillatory pitching motion was prescribed. In this stage, a dual time stepping procedure was employed and the simulation was advanced to the next step after 3 orders of magnitude convergence has been obtained.

Figure 4 shows a comparison of the pitch damping moment coefficient $(C_{m_q} + C_{m_{\dot{\alpha}}})$ between the experimental results [1], the numerical results by Vishal and Jubaraj [2], and the results obtained by the Arion code [3]. The agreement is excellent.

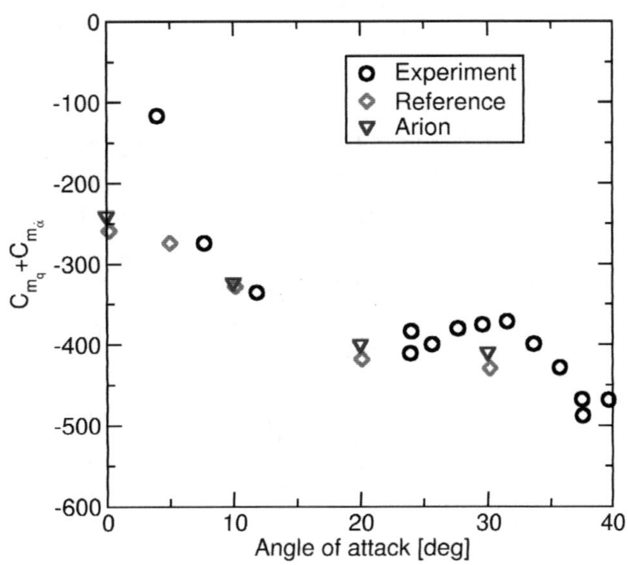

Figure 4. Pitch damping moment coefficient of forced simple pitching harmonic oscillation

III Convergence

As an example for the effect of the numerical schemes on the solution convergence, the case of $\alpha_0 = 10°$ is chosen. Three types on iterative solvers are considered. The first is based on point Gauss Seidel (PGS), the second is based on a Krylov solver [4], while the third is based on the same Krylov solver with the addition of *back-traced line search.* The following section provides a brief description of the algorithm.

III.A Back-Traced Line Search

Let Q^n be the flow field solution at iteration n and let ΔQ^n be the correction term without relaxation. The correction term is obtained by solving the system of equations that is obtained through finite volume implicit discretization, namely,

$$\left(\frac{I}{\Delta t} + J^n\right) \Delta Q^n = -R^n \tag{3}$$

where, J, the Jacobian, is given by,

$$J \triangleq \frac{\partial R}{\partial Q} \tag{4}$$

and R^n is the residual at step n. Update of the solution is conducted simply by the relation:

$$Q^{n+1} = Q^n + \alpha \Delta Q^n \tag{5}$$

where α can be used as an over- or under-relaxation parameter to enhance stability and as such, can be prescribed by the user. Alternatively, the relaxation parameter can be obtained based on the flow solution in an automatic and optimized manner. One of the methods to dynamically set the the relaxation parameter is called *back-traced line search.*

The first step of the algorithm concerns updating the solution based on Equation 3 with a relaxation parameter of unity, $\alpha = 1$. The second step is to evaluate a residual that is based on the new solution. The third step is based on fitting a second order polynomial that is based on R^n, R^{n+1}, and $J^n \Delta Q^n$. Let f be a function defined such that:

$$f(\alpha) = \frac{1}{2} R^T R \tag{6}$$

By definition,

$$\begin{aligned} f(0) &= \frac{1}{2} R^{T^n} R^n \\ f(1) &= \frac{1}{2} R^{T^{n+1}} R^{n+1} \\ f'(0) &= R^{T^n} J^n \Delta Q^n \end{aligned} \tag{7}$$

and the the second order polynomial is given by,

$$f(\alpha) = a\alpha^2 + b\alpha + c \tag{8}$$

where a, b, and c are found through the information from Equation 7. The last step is finding the value for the relaxation parameter α through a minimum of the function f. The whole process can be repeated with a third order polynomial.

III.B Steady State Simulation Convergence

As with all time-accurate flow simulations about bodies in motion, s steady state solution is sought as the initial condition of the time accurate simulation. Figure 5 shows the effect of the line search algorithm on convergence. Using line search one can reduce the computational time by 15% for 3 orders of magnitude residual drop and 35% for 4 orders of residual drop. Moreover, 6 orders of magnitude residual convergence cannot be obtained unless the line search algorithm is utilized.

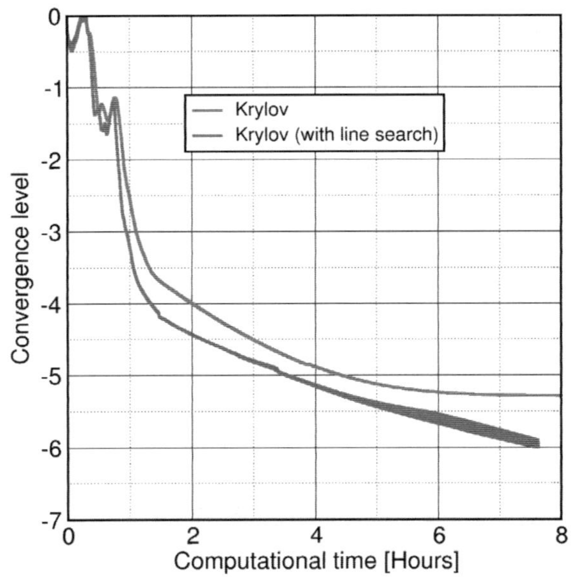

Figure 5. Effect of line search on execution time; steady state simulation

III.C Dual-Time Stepping Convergence

Dual-time stepping simulations require to prescribe two time steps. The first, the physical time step, is set based on the physics of the problem at hand and the second, a virtual time step for the iterations toward convergence. Unlike non-time-accurate simulations where the CFL number is chosen as high as possible, the CFL number in dual-time stepping simulation is effective only at a narrow range. A very high CFL number or a very low CFL number are totally ineffective. The use also must prescribe the convergence level within each physical time step.

Normally, the choice of the CFL number within a time step is chosen by the user based on experience. In the current case it was found that a 4 orders of magnitude of residual drop is necessary to obtain a sufficiently accurate moment coefficient. Therefore the convergence of the iterations is critical in terms of computational efficiency. Figure 6 shows a comparison of the computer time between all of the methods considered in the current work. The results that are shown correspond to more than two cycles of motion to make sure that the reduction is consistent. Use of the Krylov solver is found to be superior to the use of the PGS iterations. Furthermore, the use of line search further improves the convergence rates and thus reduced the overall computational time. As shown in the figure, the reduction of computational time can be as high as 40%.

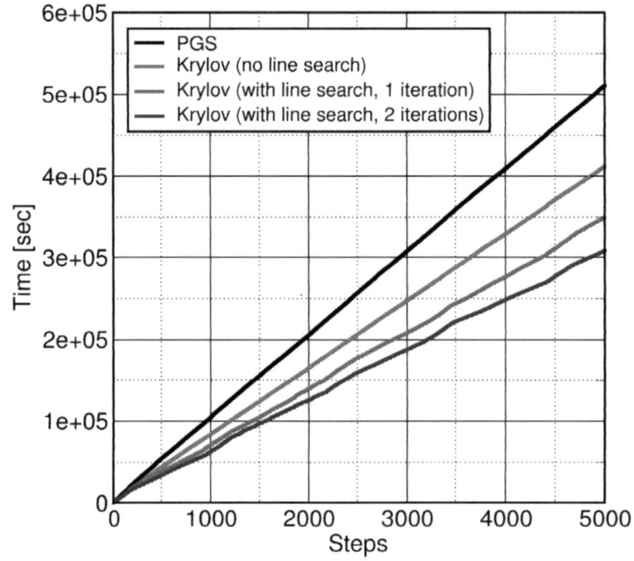

Figure 6. Effect of line search on execution time; time accurate simulations

IV Summary

The results that are shown in the current work show that the Arion can be efficiently and accurately used for the purpose of damping moment coefficient prediction. The results also demonstrate the superiority of the Krylov solver in conjunction with the back-traced line search algorithm with respect to the traditional point Gauss-Seidel iterative scheme.

References

[1] Uselton, B. L. and Uselton, J. C., "Test Mechanism for Measuring Pitch-Damping Derivatives of Missile Configurations at High Angles of Attack," *Arnold Engineering Development Center TR-AEDC-TR-75-43, Arnold AFB, TN.*

[2] Bhagwandin, V. and Sahu, J., "Numerical prediction of pitch damping derivatives for a finned projectile at angles of attack," in "50th AIAA Aerospace Sciences Meeting including the New Horizons Forum and Aerospace Exposition," , 2012, p. 691.

[3] Kidron, Y. and Levy, Y., "The Arion Unstructured Grids CFD Solver: Theoretical and User's Manual," Tech. Rep. 2023-06, Israeli CFD Center, Caesarea, Israel, 2023.

[4] Wasserman, M., Mor-Yossef, Y., Kidron, Y., and Levy, Y., "A Preconditioned Krylov Method for Turbulent Flow Simulations on Unstructured Grids," in "61st Israel Annual Conference on Aerospace Sciences," Tel Aviv, Israel, 2022.

Energy-efficient Application of Suction and Pulsed Blowing as an Alternative to a Leading Edge Slat

Shay Monat[*1]**, Asaf Kor**[2]**, Avraham Seifert**[3]**, Oksana Stalnov**[4]

[1,2,3]School of Mechanical Engineering, Faculty of Engineering, Tel-Aviv University, Israel
[4]Faculty of Aerospace Engineering, Technion — Israel Institute of Technology

*contact: Shaymonat@gmail.com

Traditional high-lift devices, such as flaps and slats, are useful to enhance aerodynamic performance. However, more innovative solutions are required to address new challenges the aviation industry faces to reach higher lift capabilities. Active Flow Control (AFC) has been studied for many decades but has not yet been applied to improve the efficiency of complex high-lift systems (Cattafesta III and Sheplak, 2011).

With new design methods, this experimental study aims to investigate the application of novel concept to a high-lift wing configuration. Herein AFC actuators that generate suction and oscillatory blowing (SaOB) (Arwatz *et al.*, 2008) are incorporated into a thick airfoil's main element to investigate the possibility of replacing the slat (see Figure 1). The benefits of having an AFC system that acts as a virtual slat is not only discussed quantitatively, with experimental results, but also practically as an efficient solution for the design of a lighter and with higher aerodynamic performance. This research characterizes the application of AFC to a wing in terms of aerodynamics and energy efficiency. The experiments were conducted in the Tel-Aviv University's sub-sonic wind tunnel (see Figure 2) at Reynolds numbers of about 1 million.

The measurements of the wind-tunnel campaign (see Figure 3) are ultimately used to evaluate the overall efficiency of the AFC system in comparison to a standard slat in three-element configuration (see Figure 1). This efficiency of the AFC system is defined using the aerodynamic Figure of Merit (FM). The FM concept was introduced by Seifert (Seifert, 2007), who suggested to consider the actuator's weight and volume as well as system complexity. Herein the FM is described as a ratio of aerodynamic performance in two- and three-element configuration,

$$FM = \frac{\left(\dfrac{C_l^{max}}{C_d + C_E/\eta}\right)^{bi}_{AFC}}{(C_l^{max}/C_d)^{tri}_{bsln}} \quad (1)$$

where C_l^{max} is the maximum lift coefficient, C_d is drag coefficient at the corresponding angle of attack, the superscripts "tri" and "bi" refer to three- and two- elements configuration in the wing (with and

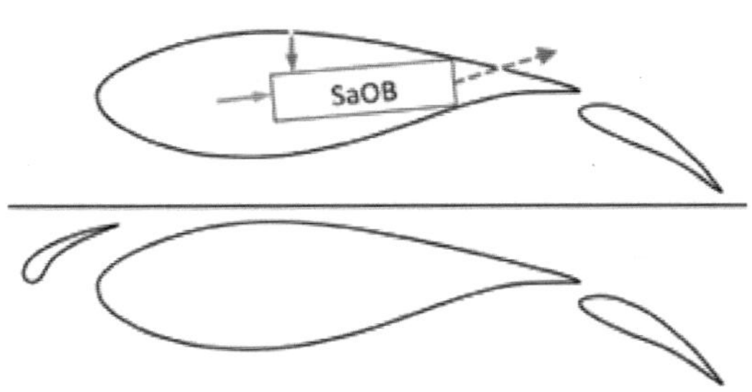

Figure 1: A conceptual sketch of the two-element wing with the SaOB actuators (top); and benchmark three-element baseline configuration (bottom).

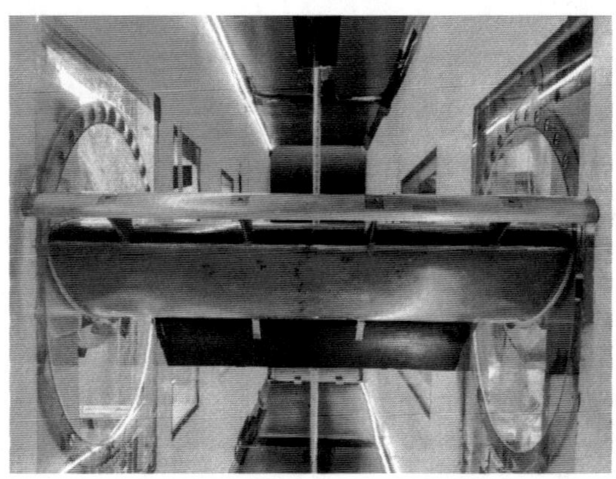

Figure 2: Photograph of the wing model in its three-element configuration installed in the wind-tunnel. Picture taken facing downstream.

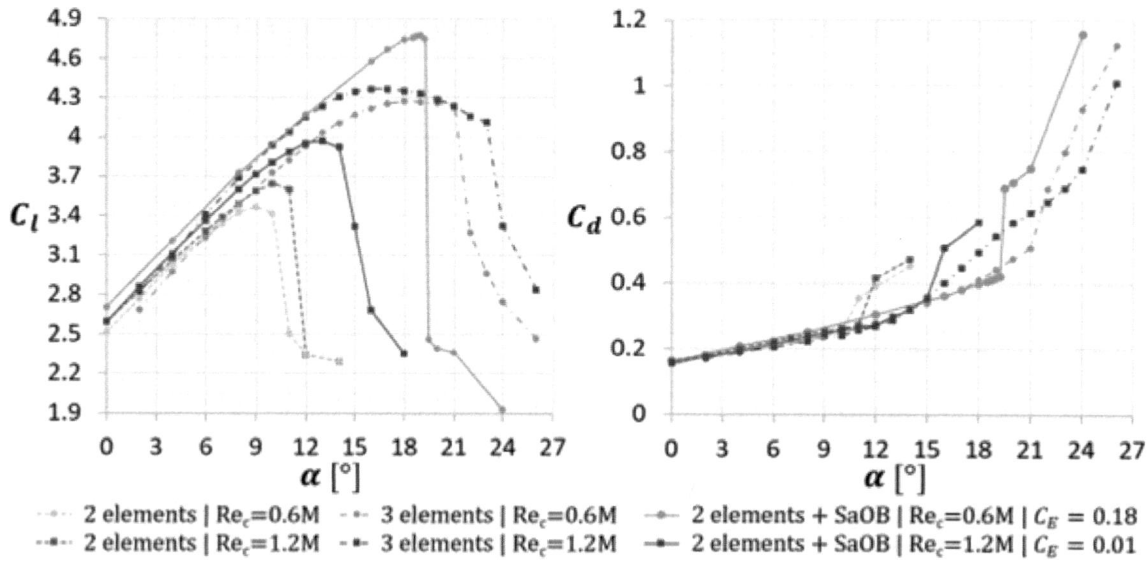

Figure 3: The lift and drag coefficients of the two-element (flap and main element), three-element (flap, main element, and slat), and two-element with SaOB configurations.

without slat), respectively. Herein η is the efficiency of any system which powers the actuators (pumps, compressors, tubes, etc.). Herein C_E is the normalized energy used to operate the AFC system:

$$C_E = \frac{P}{q_\infty U_\infty bc} \tag{2}$$

where P is the measured power input to the actuators in the two-element configuration, q_∞ is the free-stream dynamic pressure, U_∞ is the free-stream velocity, b is the span length of the wing ($608mm$) and c is the chord length of the main element ($480mm$), which is also utilized for the normalization of aerodynamic parameters (i.e. lift and drag coefficients).

The results of this study (see Figure 4) show that the AFC could replace the slat system and generate greater lift, less drag, and remain energetically efficient. A discussion on the structure of a slat-less wing with an AFC system raises the possibility that such a wing would weigh less than a wing with a slat. However, for such a system to be operational in commercial airplanes, additional challenges and tasks should be considered, such as complying with traditional safety regulations, development costs, production design, maintenance, and manufacturing challenges.

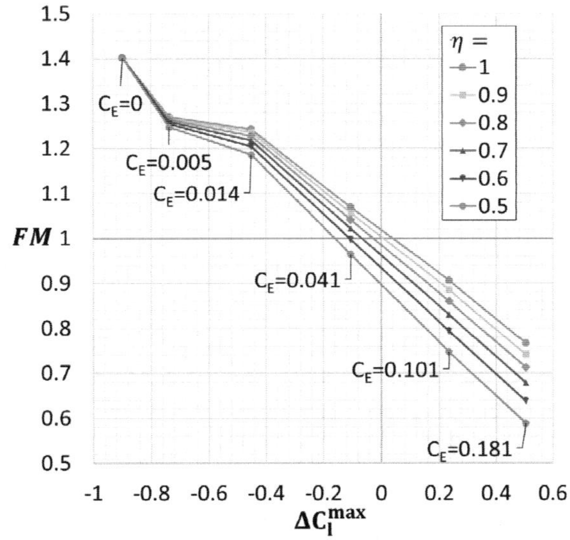

Figure 4: The figure of merit (FM) of the wing with the SaOB actuator as a function of the C_l^{max} for various values of the AFC system's efficiency (η) at $Re_c = 0.6 \times 10^6$

Acknowledgment

This research was supported by the Israeli Innovation Authority as part of the Nofar program (grant agreement number 77160).

References

Arwatz G., Fono I., Seifert A. (2008). Suction and oscillatory blowing actuator modeling and validation. *AIAA journal 46*(5), 1107–1117.

Cattafesta III L. N., Sheplak M. (2011). Actuators for active flow control. *Annual Review of Fluid Mechanics 43*, 247–272.

Seifert A. (2007). Closed-loop active flow control systems: actuators. In *Active Flow Control: Papers contributed to the Conference "Active Flow Control 2006", Berlin, Germany, September 27 to 29, 2006*, pp. 85–102. Springer.

Closed-Loop Stall Control using DBD-Plasma-Actuation

Mordechai Garcia and David Greenblatt
Faculty of Mechanical Engineering, Technion – Israel Institute of Technology

Flight endurance is a major parameter to maximized when designing the wings of unmanned aerial vehicles. For UAVs with a combination of large aspect ratios and large base drag (e.g., due to external payloads), a high lift coefficient, close to the stall angle, is required to obtain maximum endurance.[1] In order to ensure safe flight, a stall control system must be implemented In our recent wind tunnel and flight experiments, we demonstrated effective open-loop control using dielectric barrier discharge (DBD) plasma actuation on on a mini-UAV-sized RC plane model. The objective of the research at present is to implement a closed-loop scheme that will allow the DBD actuators to operate autonomously as part of a flight control system.

The sensing method used for closed-loop control is based on unsteady pressure measurements just below the leading-edge. This is essentially a simple adaptation of well-known stall-warning systems that use low pressure below the leading-edge to activate a stall warning horn or some other audible device. In order to implement the system, a leading-edge segment with five leading-edge pressure sensors was printed and adapted to an FMS Ranger PNP RC model wing (span = 1.8 m), which was tested in a 1 m × 2 m wind tunnel with a top speed of 16 m/s (Figure 1, left). The wing was equipped with a leading-edge plasma actuator and cantilever-mounted horizontally employing a six-component ATI Delta SI-330-30 balance, outboard of an endplate, at 10 cm from the tunnel wall. A stepper motor with a geared belt drive was used to attain quasi-steady and dynamic pitching of the wing at constant wind speeds of 8.5 m/s and 12.0 m/s, which corresponded to stall speeds at two anticipated wing loadings.

A sample of quasi-steady lift coefficient, drag coefficient and two leading-edge pressure port measurements is shown in Figure 1 (right). Depending on the location of the pressure ports, notice how pressure approaches zero and becomes negative as the stall angle is approached. The pressure port furthest upstream (p_1) crosses zero at $\alpha = 14°$, which corresponds to the onset of inboard trailing edge stall (observed with tufts). The pressure port slightly further downstream (p_2), crosses zero just prior to leading-edge stall. In principle, either one of these zero-crossings can be used as a sensor in the feedback loop. In the conference presentation we will demonstrate both quasi-steady and unsteady closed-loop control using these sensors.

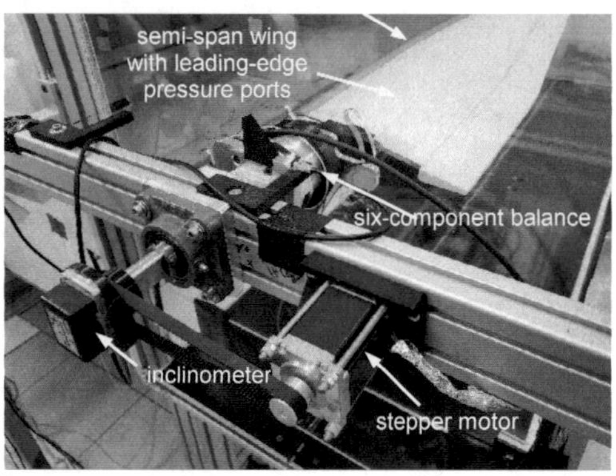

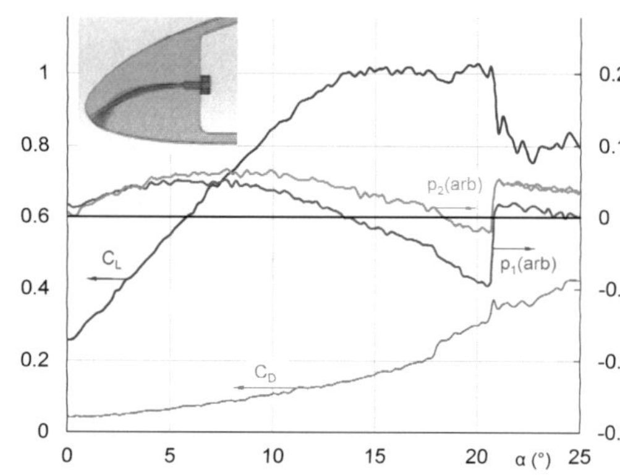

Figure 1. Photograph of the experimental setup (left); load coefficients and leading-edge pressure measurements, (right) with with pressure port locations (inset).

---

[1] Garcia, M. And Greenblatt, D., "Mini UAV with DBD-Plasma-Based Flow Control," AIAA Paper No. 2023-4306, AIAA Aviation 2023 Forum, 12-16 June 2023, San Diego, CA. https://doi.org/10.2514/6.2023-4306.

Computational and Experimental Lift Enhancement Study Applied to Wind Turbine Airfoil

Asaf Kor[1] **and Avraham Seifert**[2]

[1,2]School of Mechanical Engineering, Faculty of Engineering, Tel-Aviv University, Israel

*contact: asafkor47@gmail.com

One of the most important aspects of a wind turbine blade performance is its lift coefficient (C_L); the higher the C_L, the more efficient is the wind turbine's capability of converting wind power to electric power. This study focuses on the improvement of the maximum C_L of a wind turbine blade profile (C_L) using active flow control devices (PFC and AFC) Bianchini *et al.* (2019). Computational fluid dynamics (CFD) analysis was carried out to determine the optimal geometry which included the main element, a leading edge slat and a trailing edge flap. Figure 1 (a) and (b) show the different slat positions used to select the best performing one. The CFD results were compared to wind tunnel (WT) experiments, where a series of experiments on several different aspect ratio wings that were 3D printed.

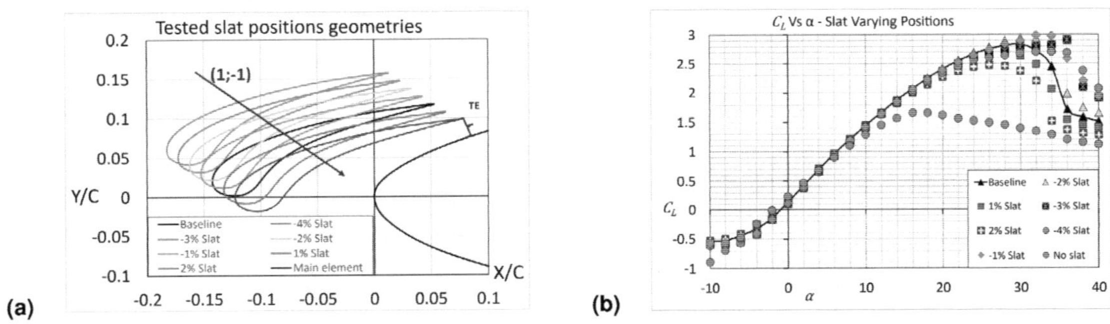

Figure 1: Slat optimization process. The slat was translated from an original position along the line shown. (a) Shows the tested slat locations, and (b) shows the lift coefficients of the different positions.

Figure 2 shows the experimental and simulation results, which became the baseline results for the subsequent AFC simulations and WT experiments. At first, the AFC was simulated with Reynolds average Navier-Stokes (RANS) model, which was also used in the earlier stages Crivellini and D'Alessandro (2014); Saraf *et al.* (2012). The results then showed desired possible improvement in the C_L when steady suction was applied on the slat's leading edge region, at high angles of attack (AOAs). However, the RANS simulations were too optimistic and failed to consistently detect flow separation on any of the elements at stall. This led to predict that suction on the slat will have a minor effect. Transient AFC simulations in the form of steady suction using a 2D adapted large eddy simulations (LES) turbulent model was then used as a better model for detecting flow separation and its control as shown in Figures 3 (a) and (b). Indeed, the LES simulations were able to detect boundary layer (BL) separation and provide a more accurate prediction of slat suction effectiveness. Application of steady suction on the slat showed BL separation delay, along with a small improvement in C_L. Finally, another round of WT experiments were performed in order to validate the results of the AFC simulations. The results are shown in Figure 4 and demonstrate that the LES model is more suitable in predicting the effects of AFC activation, compared to RANS.

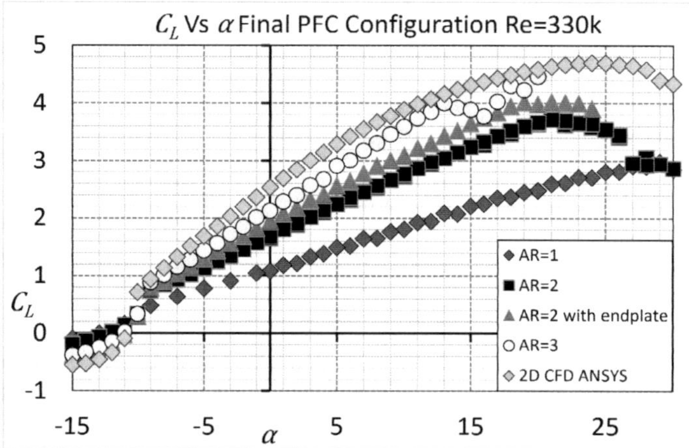

Figure 2: WT experiment of the PFC wing, compared to the CFD results.

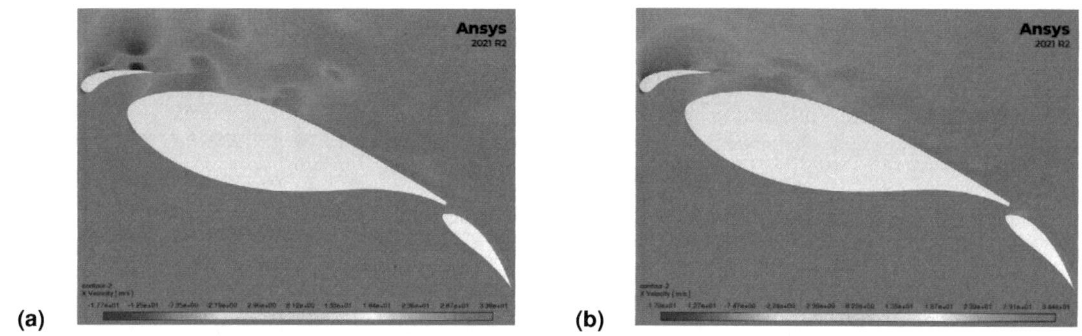

Figure 3: Velocity contours created in ANSYS to show flow separation on the elements using transient LES model. (a) Without AFC (b) with suction from the slat's LE.

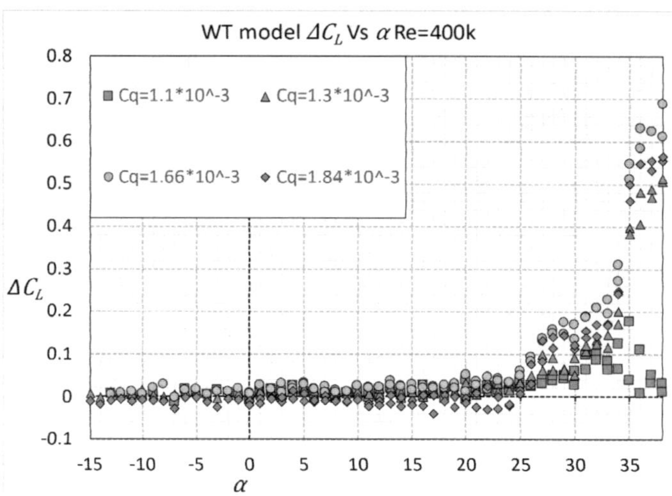

Figure 4: Lift coefficient difference plot of the WT experiment results of the wing with AFC, at different suction coefficient C_q values, compared to the baseline.

References

Bianchini A., Balduzzi F., Di Rosa D., Ferrara G. (2019). On the use of gurney flaps for the aerodynamic performance augmentation of darrieus wind turbines. *Energy conversion and management 184*, 402–415.

Crivellini A., D'Alessandro V. (2014). Spalart–allmaras model apparent transition and rans simulations of laminar separation bubbles on airfoils. *International Journal of Heat and Fluid Flow 47*, 70–83.

Saraf A. K., Singh M., Kumar A. (2012). Analysis of the spalart-allmaras and kmω standard models for the simulation of the flow over a national advisory committee for aeronautics (naca) 4412 airfoil.

Abstract for the 63-rd IACAS 2024 Meeting

Sensitivity of flows over three-dimensional swept wings at low Reynolds number

Anton Burtsev*
University of Texas at Austin, Austin, TX 78712, US
Vojtech Pezlar†
University of Liverpool, Liverpool L69 3GH, UK
and Vassilis Theofilis‡
Technion - Israel Institute of Tehcnology, Haifa, Israel

High-angle of attack flows over swept three-dimensional wings based on the NACA 0015 profile are studied numerically at low Reynolds numbers. Linear stability analysis is used to compute instability and receptivity of the flow via the respective three-dimensional (TriGlobal) direct and adjoint eigenmodes. Leading TriGlobal modes ($\hat{\boldsymbol{u}}$) are compared with earlier work on untapered wings.[1] The magnitude of the adjoint eigenvectors is used to identify regions of maximum flow receptivity to momentum forcing. It is found that such regions are located above the primary three-dimensional separation line, their spanwise position varying with wing sweep. The wavemaker region corresponding to the leading global eigenmode is computed and found to lie inside the laminar separation bubble (LSB) at the spanwise location of peak recirculation. Increasing the Reynolds number leads to the wavemaker becoming more compact in the spanwise direction and concentrated in the top and bottom shear layers of the LSB. As sweep is introduced, the wavemaker moves towards the wing tip, following the spanwise displacement of maximum recirculation. Flow modifications resulting from application of different types of forcing are studied by direct numerical simulation initialized with insights gained from stability analysis. Steady suction applied at the location of maximum receptivity shows a reduction of the amplitude of lift and drag oscillations.

*Postdoctoral Scholar, Departement of Aerospace Engineering and Engineering mechanics, UTA
†Doctoral Student, Faculty of Fluid Mechanics and Thermodynamics, CTU
‡Professor, Departement of Aerospace Engineering, Technion

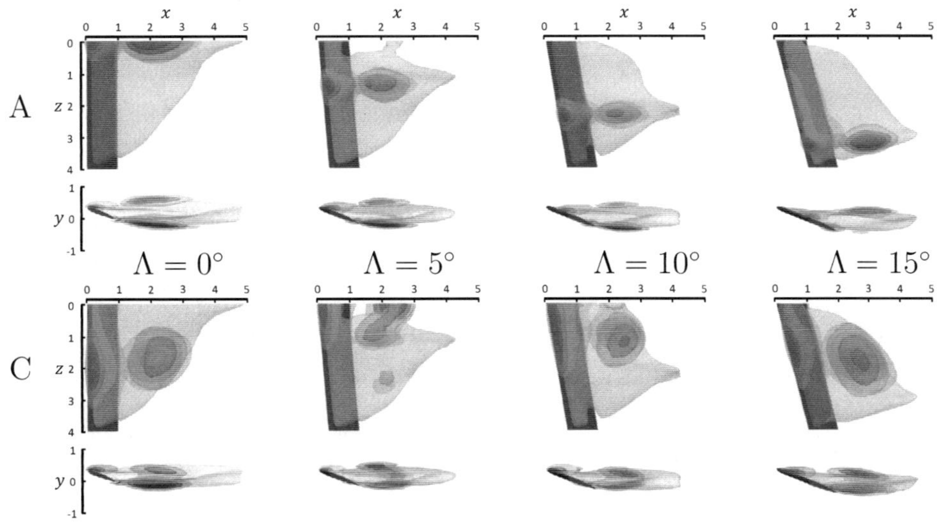

Figure 1. The effect of sweep angle (Λ) on the structural sensitivity $\|\hat{\boldsymbol{u}}^{+}\|\cdot\|\hat{\boldsymbol{u}}\|$ (orange) and receptivity to momentum forcing $\|\hat{\boldsymbol{u}}^{+}\|$ (blue) of modes A and C.

Adjoint TriGlobal eigenmodes on a finite aspect ratio wing at high angle of attack have been computed for the first time. The regions of maximum sensitivity and receptivity were calculated, and their variation with Reynolds number and wing sweep angle was documented in the low-Re range examined.

The wavemaker of the leading global eigenmode was shown to lie inside the LSB at the spanwise location of peak recirculation. At low reynolds numbers, it is qualitatively similar to that of spanwise homogeneous wings.[3] Increasing Re leads to the structures of both the global mode and the wavemaker becoming more compact in the spanwise direction. At the highest presently considered $Re = 500$, the wavemaker is concentrated in the top and bottom shear layers of the LSB. As sweep is introduced to the wing, the wavemaker moves towards the wing tip following the spanwise displacement of maximum recirculation see Fig. 1. The region of maximum receptivity of the leading eigenmode is located near the separation line and is compact in the spanwise direction. Its spanwise position is linked to the reattachment node present on the suction side of the wing in the steady base flow. The compact nature of both the wavemaker and the regions of maximum receptivity to momentum forcing and the variation of their locations with increased sweep angle is of practical significance for flow control. This implies that if actuators were positioned evenly along the span of the wing, only a few that coincide with the location of the receptivity region would be most effective.

Modification of the flow by periodic and steady forcing was applied at locations of maximum receptivity to momentum forcing identified using the stability analysis. Periodic forcing at the frequency of the leading global mode was shown to accelerate the onset of wake shedding. Steady blowing showed a reduction of lift to drag ration (C_L/C_D), whereas steady suction reduced the amplitude of C_L/C_D oscillations.

References

[1] A. Burtsev, W. He, K. Zhang, V. Theofilis, K. Taira, and M. Amitay. Linear modal instabilities around post-stall swept finite wings at low Reynolds numbers *J. Fluid Mech.*, 944, A6, 2022.

[2] F. Giannetti, and P. Luchini. Structural sensitivity of the first instability of the cylinder wake. *J. Fluid Mech.*, 581, 167–197, 2007.

[3] G. Nastro, J.-C. Robinet, J.-C. Loiseau, P.-Y. Passaggia, and N. Mazellier. Global stability, sensitivity and passive control of low-Reynolds-number flows around NACA 4412 swept wings. *J. Fluid Mech.*, 957, A5, 2023.

Abstract for the 63-rd IACAS 2024 Meeting

Linear Stability of Compressible Flows in Open-Cavities with Curved Downstream Walls

Vojtech Pezlar*

Czech Technical University, Prague 160 00, Czech Republic

Kamil Dylewicz†

University of Liverpool, Liverpool L69 3GH, UK

and Vassilis Theofilis‡

Technion - Israel Institute of Tehcnology, Haifa, Israel

Linear three-dimensional stability analysis is performed in subsonic and supersonic open cavity flows. An S-shaped downstream wall is mapped analytically onto a rectangular computational domain and a novel multi-domain method is developed to analyze modal stability in the sub- and supersonic regimes. Results show attenuation of unsteadiness for cavities with modified downstream wall compared to their rectangular counterparts.

The present study follows upon the experimental findings of Das and Cohen[1],[2] who showed attenuation of unsteadiness of transonic flows over open cavities via modification of downstream cavity wall (to circular and elliptic shapes). The obtained effect, which can be used to mitigate unwanted aerodynamical phenomena and pressure oscillations present on aerospace structures, was attributed to the disruption of the Rossiter resonance mechanism. In turn, global linear stability analyses have associated the Rossiter mechanism with the leading unstable mode of the shear-layer formed at the open end of the cavity.[3]

Presently, linear stability analysis of nominally two-dimensional laminar steady states in cavities with analytically-defined curvilinear boundaries is performed. To this end, a

*Doctoral Student, Faculty of Fluid Mechanics and Thermodynamics, CTU

†Doctoral Student, School of Engineering, UoL

‡Professor, Departement of Aerospace Engineering, Technion

novel multi-domain spatial discretization method was developed to address curved geometry boundaries and was applied to define an S-shaped downstream cavity wall. The resulting compressible, non-symmetric, generalized eigenvalue problem pertinent to BiGlobal stability analysis[4] was discretized using compact high-order finite-difference scheme and solved using Arnoldi iteration.

Validation was performed against the work of Bres and Colonius[5] in the subsonic regime, and that of Sun *et al.*[3] in supersonic flow. Figs. 1a and 1b show the 2D steady laminar basic flows at the free-stream conditions studied in.[5] Figs. 1c and 1d confirm the integrity of our numerical solutions in the case of rectangular downstream cavity wall, while Figs. 1e and 1f show the spanwise component of the leading eigenvalues of the rectangular and S-shaped geometries. Stability analysis of the quarter-elliptical open-cavity geometry studied by Das and Cohen[12] is underway and results will be presented at the time of the conference. Future work will extend the stability analysis of the S-shaped cavity geometry into the hypersonic flow regime and present comparisons with ongoing experiments.

References

[1]Das, S. and Cohen, J., "Reducing Pressure Oscillations by Change in Rear-Face Geometries at Subsonic and Transonic Speeds," *AIAA J.*, Vol. 55, No. 12, 2017, pp. 4401–4411.

[2]Das, S. and Cohen, J., "Reducing Cavity Flow Pressure Oscillations by Change in Rear Face Geometries at Transonic Speed," 21st Australasian Fluid Mechanics Conference, Adelaide, Australia, December 2018.

[3]Sun, Y., Taira, K., Cattafesta, L., and Ukeiley, L., "Biglobal instabilities of compressible open-cavity flows," *J. Fluid Mech.*, Vol. 826, 2017, pp. 270–301.

[4]Theofilis, V., "Advances in global instability of non-paralle and three-dimensional flows," *Prog. Aero. Sci.*, Vol. 39, 2003, pp. 249–315.

[5]Bres, G. and Colonius, T., "Three-dimensional instabilities in compressible flow over open cavities," *J. Fluid Mech.*, Vol. 599, 2008, pp. 309–339.

(a) Streamwise velocity component of the base flow at $Re = 2110, Ma = 0.6$[5]

(b) Same as in Fig. 1 (a) but with the S-shaped downstream wall

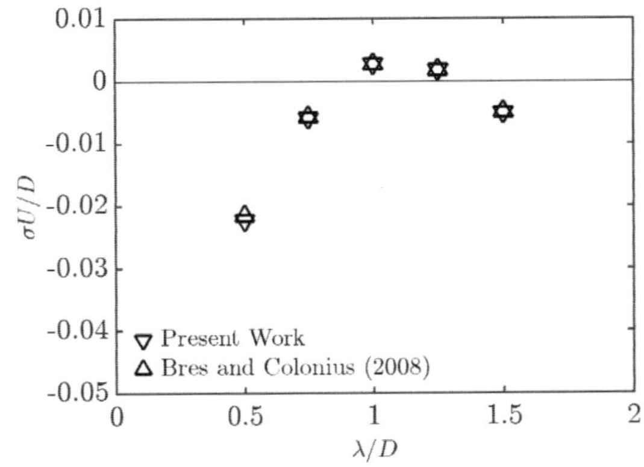

(c) Comparison with[5] of scaled amplification rate, σ, as a function of the spanwise periodicity length λ of the leading eigenmode.

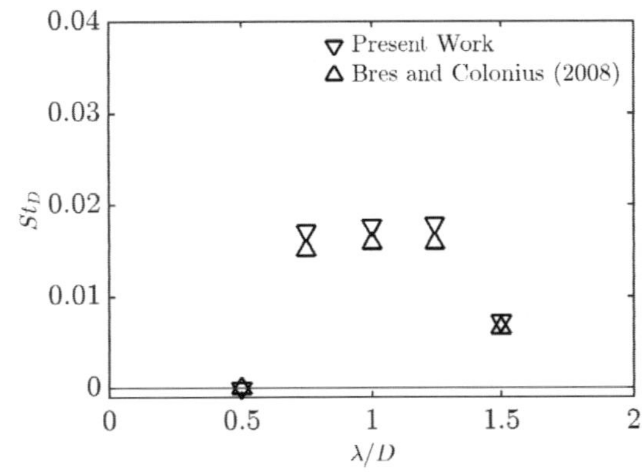

(d) Comparison with[5] of the scaled circular frequency, $St_D = \omega D / 2\pi U_\infty$, as a function of the spanwise periodicity length λ of the leading eigenmode.

(e) Leading spanwise velocity perturbation ($\Re\{\hat{w}\}$) of the rectangular wall configuration

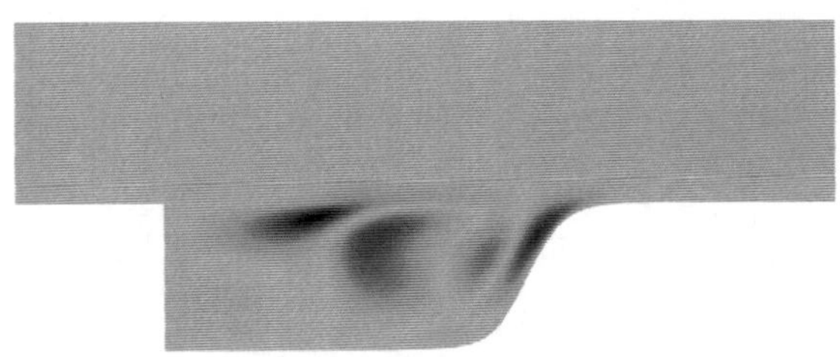

(f) Leading spanwise velocity perturbation ($\Re\{\hat{w}\}$) of the S-shaped wall configuration

Figure 1: Comparison of the effect of downstream cavity wall modification on linear flow stability. *Left Column:* Rectangular wall; *Right Column:* S-shaped wall.

Abstract for the 63-rd IACAS 2024 Meeting

Investigation of global instabilities on ROTEX-T cone-flare geometry

Kamil Dylewicz*
University of Liverpool, Liverpool L69 3GH, UK
Vassilis Theofilis†
Technion - Israel Institute of Tehcnology, Haifa, Israel

Viscous laminar Mach 6 flow over a cone-flare geometry is investigated numerically to determine the critical Reynolds number past which self-excited instabilities inside the boundary layer naturally lead flow to laminar-turbulent transition. To this end, direct numerical simulations (DNS) are performed to converge the unsteady axisymmetric equations of motion to steady-state solutions. Subsequently, global linear modal and non-modal stability analyses are performed, taking the azimuthal direction to be homogeneous, to unravel the leading unstable three-dimensional (non-axisymmetric) perturbations. Finally, results are verified in full DNS which follows the linear evolution of the perturbations up to the early nonlinear breakdown.

I. Introduction

The present work investigates the cone-flare ROTEX-T[1] configuration with respect to its linear, modal and non-modal, perturbations that lead axisymmetric flow to laminar-turbulent transition. These small-amplitude disturbances are associated with shock wave/boundary layer interactions (SBLIs), as well as amplified and self-excited perturbations of the laminar

*Doctoral Student, School of Engineering, UoL
†Professor, Departement of Aerospace Engineering, Technion

separation bubble (LSB) formed at the cone-flare. Herein, the flow conditions corresponding to a vehicle cruising at altitude of $50km$ according to U. S. Standard Atmosphere, with free-stream Mach number $M_\infty = 6.0$, were selected in order to maintain laminar and steady flow over the geometry while achieving the size of the model that is of practical relevance. Additionally, such choice of flow conditions ensures that the second Mack mode,[2] which develops inside the boundary layer prior to the flare, is convectively stable, which permits neglecting interaction of boundary layer instabilities with those developing in the flare region on account of SBLI.

II. Ongoing work

Our recent work at high unit Reynolds numbers[3] provided evidence of global instability of the axisymmetric flow. Consequently, here we focus on the identification of conditions at which the laminar flow is truly steady and axisymmetric on the cone preceding the flare, and compute this flow in a time-accurate manner until a true steady-state has been reached. One of the objectives of the present investigation is the identification of a critical Reynolds number at which the flow becomes globally unstable on account of modal linear perturbations.

To this end, a steady-state is obtained by means of a direct numerical simulations over the ROTEX-T geometry shown in figure 1 (left) where $x_1 = 0.4377m$, $x_2 = 0.55m$, $\alpha = 7°$, $\beta = 20°$ and $Re_\xi = \rho_\infty u_\infty x_1/(\mu_\infty \cos\alpha) = 5.0 \times 10^4$. Here, an open-source *rhoCentralFoam* code[4] has been used with second-order temporal scheme and CFL number maintained below 0.75 throughout the simulation. Time-accurate convergence of the flow towards the steady-state has been verified using residuals computed within the boundary layer using probes; for their approximate locations see p_1, p_2, p_3 and p_4 in figure 1 (left). The steady-state solution is shown, using Mach number field, in figure 1 (right). Residuals convergence proves that the flow remains globally stable in the axisymmetric sense. However, previous experience on this geometry[3] might suggests that the onset of the global instability happens on account of three-dimensional perturbations, sustained within the separated shear layer, with high periodicity along the azimuthal direction. Therefore, presently computed steady-state is a good candidate to begin systematic investigations of modal and non-modal three-dimensional linear perturbations to eventually determine critical flow conditions past which self-excited instabilities inside the boundary layer naturally lead flow to laminar-turbulent transition.

III. Future work

Future work includes mesh independence study of the steady-state solution at present Reynolds number of 5.0×10^4 followed by modal and non-modal global stability analysis

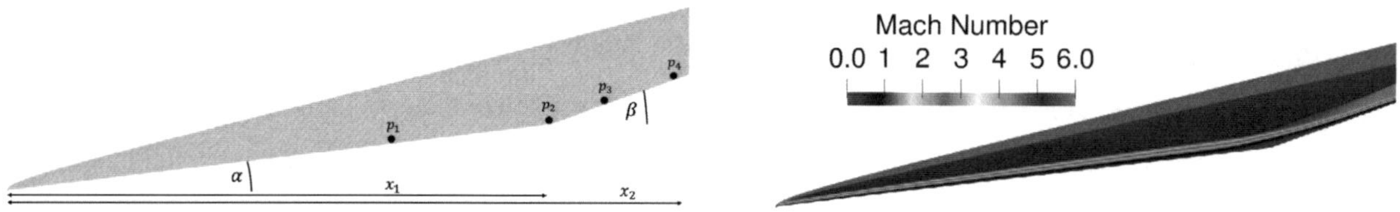

Figure 1: Geometry outline with chosen parameters and probes' locations indicated (left) and Mach number flow field for a converged steady-state at $Re_\xi = \rho_\infty u_\infty x_1/(\mu_\infty \cos\alpha) = 5.0 \times 10^4$.

using extensively validated BiGlobal stability code.[5]

Acknowledgement

This work is supported by the AFOSR under award number FA8655-23-1-7031 with Dr. Douglas Smith as PO. Computational time provided by UKTC and EPSRC on the UK supercomputing facility ARCHER2 via projects EP/R029326/1 and EP/X035484/1 is gratefully acknowledged.

References

[1]Gülhan, A., Thiele, T., Siebe, F., Klingenberg, F., Kronen, R., Stamminger, A., Scheuerpflug, F., Kallenbach, A., and Jung, W., "Main Achievements of the Rocket Technology Flight Experiment ROTEX-T," 21st AIAA International Space Planes and Hypersonics Technologies Conference, 2017.

[2]Mack, L. M., "Boundary-layer linear stability theory," Tech. rep., California Inst of Tech Pasadena Jet Propulsion Lab, 1984.

[3]Davami, J., Juliano, T. J., Scholten, A., Paredes, P., Benitez, E. K., Running, C. L., Dylewicz, K., Pezlar, V., Theofilis, V., Thiele, T., et al., "Separation and Transition on the ROTEX-T Cone-Flare," *AIAA SCITECH 2024 Forum*, 2024, p. 0499.

[4]Greenshields, C. J., Weller, H. G., Gasparini, L., and Reese, J. M., "Implementation of semi-discrete, non-staggered central schemes in a colocated, polyhedral, finite volume framework, for high-speed viscous flows," *International journal for numerical methods in fluids*, Vol. 63, No. 1, 2010, pp. 1–21.

[5]Quintanilha, H., Paredes, P., Hanifi, A., and Theofilis, V., "Transient growth analysis of hypersonic flow over an elliptic cone," *Journal of Fluid Mechanics*, Vol. 935, 2022, pp. A40.

Kinetic description of flow detachment at a micro-step

D. Ben-Adva,[1, a)] G.Tatsios,[2] and A. Manela[1]

[1)] *Faculty of Aerospace Engineering Technion Haifa 32000 Israel*

[2)] *School of Engineering Institute for Multiscale Thermofluids University of Edinburgh Edinburgh EH9 3FB UK*

We study the pressure-driven steady gas flow, imposed by temperature or density gradients, over a backward-facing step in a two-dimensional microchannel. Focusing on the near-free-molecular regime of high Knudsen (Kn) numbers, the problem is analyzed asymptotically based on the Bhatnager, Gross and Krook kinetic model, and supported by numerical Discrete Velocity and Direct Simulation Monte Carlo calculations. The wall conditions are formulated using the Maxwell model, superposing specular and diffuse surface models. The asymptotic solution contains the leading-order free-molecular description and a first-order integral representation of the near-free-molecular correction. In contrast with the common view that flow detachment is a low-Knudsen-number phenomenon [1, 2, 3], and in line with recent works indicating the occurrence of detachment at high Knudsen numbers [4, 5], our results show that flow separation at the step may occur at arbitrarily large (yet finite) Knudsen numbers in sufficiently smooth channels, as illustrated in Fig. 1. It is then shown that detachment is significantly suppressed by the imposition of density variations between the reservoirs and an increase in the channel walls accommodation coefficient towards diffuse-surface conditions. While the mass flow rate in a specular channel decreases with decreasing $\mathrm{Kn} \gg 1$ in a density-driven setup (in line with the Knudsen Paradox), it increases in a temperature-driven flow. The results are obtained for arbitrary differences between the inlet and outlet reservoir equilibrium properties, and are rationalized using the linearized problem formulation.

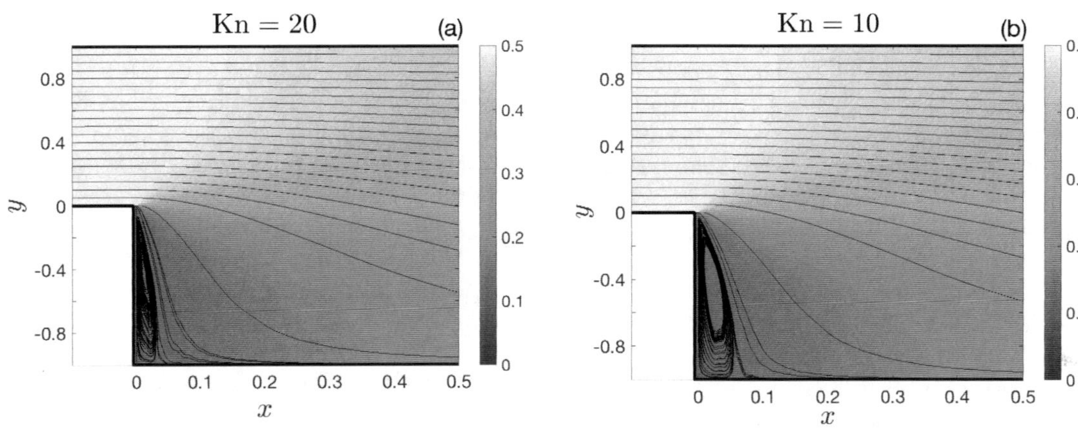

Fig. 1. Effect of the Knudsen number on the flow streamlines (curves) and pressure field (colormaps) for a temperature-driven flow in a specular-wall channel. Results are presented at the indicated inlet-channel-based Knudsen numbers of Kn=20 (Fig. 1a) and Kn=10 (Fig. 1b).

ACKNOWLEDGMENTS

The research was supported by the Israel Science Foundation (grant No. 412/21).

REFERENCES

[1]A. Beskok, "Validation of a new velocity-slip model for separated gas micro flows," Numerical Heat Transfer B **40**, 451–471 (2001).
[2]O. Baysal, N. Erbas, and M. Koklu, "Control of separated flow past a backward facing step in a microchannel," Microfluidics and Nanofluidics **1**, 86–92 (2004).
[3]F. Bao and J. Lin, "Continuum simulation of the microscale backward-facing step flow in a transition regime," Numerical Heat Transfer A **59**, 616–632 (2011).
[4]A. Manela and L. Gibelli, "Free-molecular and near-free-molecular gas flows over backward facing steps," Journal of Fluid Mechanics **889**, A22 (2020).
[5]A. Mahdavi and E. Roohi, "A study on micro-step flow using a hybrid direct simulation Monte Carlo-Fokker-Planck approach," Physics of Fluids **34**, 062007 (2022).

a)Corresponding author: din-ben-adva@campus.technion.ac.il

DEPARTMENT OF
AEROSPACE ENGINEERING

FMLAB

Investigation of near-wall scatter of scaled inflectional mean velocity profiles of turbulent separated flow over the Gaussian bump

Eric Vaizman*, Igal Gluzman†

Technion - Israel Institute of Technology, Haifa 3200003, Israel

Velocity profile scaling in turbulent separated flows has been investigated for the available experimental data of the smooth body separation experiment over the Boeing Gaussian bump. Application of the Embedded Shear Layer (ESL) scaling that was proposed by [Schatzman and Thomas, 2017] was shown to give rise to a tight collapse of inflectional mean velocity profiles over a large wall-normal extent with a large deviation occurring near the wall region, through separation and reattachment. Herein, we focus our investigation on the near-wall scatter where a new scaling, which we denote as the Split Region ESL (SRESL), is proposed to account for two distinct regions governed by different flow parameters above and below the inflection point in each mean velocity profile along the separation bubble. With the new SRESL scaling, an improved collapse of mean velocity profiles on top of each other was obtained, resembling the sigmoid function family. However, the transformed profiles by SRESL scaling still exhibited scatter in the near wall region to some extent. To characterize the deviation of the scaled profiles near the wall, we applied a Generalized Logistic Function (GLF) fit, which can match the shape of each scaled profile, and studied the fitted parameter trends along the separation bubble range to establish their relation to separated flow characteristics. The behavior of the parameters of the GLF fit was analyzed for experimental data at two freestream Mach numbers Ma = 0.1, and 0.2. Relations were established between several GLF parameters and physical quantity measurements of the flow, such as velocity at the inflection point, local free stream velocity, and the skin friction streamwise variation in the separation bubble region. The observed relations exhibited similar trends across both Mach numbers that were examined. These established relations between GLF parameters and the flow physics have the potential of representing a wide range of velocity profiles in separated flow fields via an analytical function in a wide range of flow geometries and Reynolds numbers.

References

[Schatzman and Thomas, 2017] Schatzman, D. and Thomas, F. (2017). An experimental investigation of an unsteady adverse pressure gradient turbulent boundary layer: embedded shear layer scaling. Journal of Fluid Mechanics, 815:592–642.

*Undergraduate Student, Faculty of Aerospace Engineering. E-mail: eric.vaizman@campus.technion.ac.il
†Assistant Professor, Faculty of Aerospace Engineering. E-mail: igal.gluzman@technion.ac.il

On instabilities of entropy-layers in compressible flow

Predicting the onset of turbulence in supersonic flows is crucial due to the increased friction drag and thermal loads. Understanding how external disturbances convert into specific instability modes within boundary layers, a process known as receptivity, is important for flow control and mitigating drag and aerodynamic heating effects.

The foundational understanding of stability and transition of high-speed flows has been gained by applying the linear stability theory (LST) to supersonic and hypersonic boundary layers. These works identified inviscid instabilities which are linked to the presence of a generalized inflection point (GIP) in the flow, where GIP is defined as the point maximizing $\overline{\rho}\partial\overline{u}/\partial y$.

Our investigation focuses on instabilities associated with entropy layers in inviscid hypersonic blunt body flow. These entropy layers may possess instabilities due to the presence of a GIP. The base flow is computed using a time-dependent shock-fitting algorithm, treating the shock wave as a boundary condition. This approach enables avoiding the need to capturing the shock with artificial viscosity and has proven effective in studies of receptivity in the past.

Despite of the complexity of the base flow, our analysis concentrates on a region where the flow can be considered locally parallel, allowing for the application of local stability theory. Here, we examine a cross-section far from the leading edge where the flow aligns with the blunt body's constant geometry. Utilizing local LST, the Mach Independence principle is demonstrated for entropy-layer instabilities in hypersonic flows. Therefore, our findings imply that these instabilities can be transferred from one Mach number to another via a proper normalization. The influence of the boundary conditions at the shock is also discussed.

Drastic reduction of Cavity Flow Pressure Oscillations at Supersonic Speed by modifying its Rear Face into an Ellipse shape

Soumya R. Nanda[1,2], S. K. Karthick[1,3], Sudip Das[1,4], and Jacob Cohen[1]

[1]Faculty of Aerospace Engineering, Technion-Israel Institute of Technology, Haifa-3200003, Israel

[2]Department of Aerospace Engineering, Indian Institute of Technology Kanpur, Kanpur-208016, India

[3]Department of Mechanical and Aerospace Engineering, Indian Institute of Technology Hyderabad, Kandi-502284, India

[4]Department of Space Engineering and Rocketry, Birla Institute of Technology Mesra, Ranchi-835215, India

The present study investigates the effect of a geometric change to the cavity rear face from vertical to a wedge, circular, or elliptic shape in supersonic flows. It is demonstrated, experimentally and numerically that like previous findings reported in transonic flow [1], compared to a regular open rectangular cavity at a transonic Mach number of M=0.9, this technique immensely reduces and may even completely eliminates the self-excited strong cavity flow oscillations (Rossiter's modes). The experiments were carried out in a newly constructed hypersonic Ludwieg tunnel at a freestream design Mach number of M_∞ = 6.0. The almost two-dimensional (2D) cavity (width-to-depth ratio of 10) having a length-to-depth ratio of 3.575 is placed on a 2D symmetric wedge having a half inclination angle of 41^0, so the Mach number approaching the cavity is 3.32. The Reynolds number, based on the depth of the cavity was varied between 51,500 to 74,000. Qualitative and quantitative assessments are made using Schlieren imaging, planar laser Rayleigh scattering, and unsteady pressure measurements. As the upstream approaching flow is laminar, the complementary computations are carried out without any modeling using ANSYS-Fluent and agree with the experiments. A large amplitude discrete cavity tone was observed for the vertical rear face cavity and is well predicted by Rossiter semi-empirical formula (for the 3rd mode). The generation of the Rossiter modes is due to the cavity flow feedback loop, between the Kelvin-Helmholtz vortices propagating along the shear layer and the acoustic waves generated by the flow impact on the rear edge of the cavity. Modification of the vertical rear cavity shape to an ellipse almost completely alleviates the cavity tone.

[1] S. Das and J. Cohen. “Reducing Cavity Flow Pressure Oscillations by Change in Rear Face Geometries at Transonic Speed”, 21st Australasian Fluid Mechanics Conference Adelaide, Australia 10-13 December 2018

The Initial Stages of a Pressurized Hydrogen Jet Release: Analysis via High-Fidelity Numerical Simulations

Odie Nassar[1], Moran Ezra[1], Marcel Martins Alves[1], Sergey Kudriakov[2], Etienne Studer[2], Liel Ishay[3], and Yoram Kozak[1]

[1]*Tel Aviv University, School of Mechanical Engineering, Tel Aviv, 69978, Israel*
[2]*DES/ISAS-DM2S-STMF, CEA, Université Paris-Saclay, Gif-sur-Yvette, France*
[3]*Nuclear Research Center Negev, Negev, Israel*

High-pressure hydrogen storage is essential for both industrial and aeronautical applications, such as gas-turbine, internal combustion, and scramjet engines. Unlike traditional hydrocarbon fuels, a sudden release of pressurized hydrogen as a jet into ambient air can lead to spontaneous ignition and the formation of a jet flame. Intentional promotion or prevention of this phenomenon is possible by gaining better understanding on the highly turbulent hydrogen jet flow during the initial stages of the release. For this purpose, in the current study, we carry out high-fidelity massively parallel 3-D numerical simulations using the Implicit Large Eddy Simulation (ILES) approach. In particular, we explore the influence of the initial pressure ratio between hydrogen and atmospheric air on the turbulent jet flow characteristics and mixing. For gaining fundamental insights on the initial transient flow phenomena, we focus on conditions where spontaneous ignition is not possible. Thus, we investigate the influence of different initial pressure ratios of 60, 30, 15, and 7.5 with a 3 mm jet release diameter. First, the steady-state Mach disk positions predicted by our simulations are validated with excellent agreement against experimental results from the literature. Then, we examine via our numerical simulations the transient evolution of the Mach disk position. We reveal that the time required for the Mach disk to reach steady-state conditions decreases as the initial pressure ratio decreases. Additionally, we demonstrate that the turbulent hydrogen jet is self-similar at the early stages of the release. However, we also show that as the initial pressure ratio decreases, the jet self-similarity is lost at an earlier stage. According to these results, we develop a new generalized correlation for the jet penetration depth evolution during the initial transients. Finally, we analyze in detail the mixing shear layer induced by the turbulent jet flow. We show that as the initial pressure ratio increases, the shear mixing layer becomes thicker and exhibits a much broader temperature range. The current research outcomes are aimed to assist in the design of hydrogen injection systems for aerospace propulsion applications, and to provide improved high-pressure hydrogen storage safety regulations.

Cooperative Guidance for Simultaneous Interception Using Multiple Sliding Surfaces

Maximillian Fainkich *and Tal Shima †

A cooperative guidance strategy is developed to force multiple missiles to intercept a target simultaneously. The guidance law works to minimize the time-to-go difference between neighboring missiles while still keeping the missiles on track for interception. The guidance law is derived using sliding mode control implementing multiple different sliding surfaces, with one sliding surface for every pair of missiles to remove time-to-go difference and one global sliding surface to make sure at least one missile is heading towards the target, guaranteeing the others as well. The sliding surface between every two pairs of missiles is independent of the location or order of the missiles and does not require the missiles themselves to be next to each other. The global sliding surface is defined as the product of the rates of changes of the line of sight angles of the missiles. A time-to-go approximation scheme for a missile pursuing a stationary target using proportional navigation is used during the derivation. Stability of the guidance law given the uncertainty in the calculation of the time to go is proven using a Lyapunov analysis. In addition, the guidance law is expanded to a moving target using a predicted interception point approximation scheme. A switch to proportional navigation is implemented in the final stages of the encounter in order to guarantee interception of the target. A two-dimensional nonlinear simulation of the relative kinematics is run for cases of both two missiles and more, as well as a moving target, in which the guidance law is shown to successfully cause simultaneous interception between multiple missiles starting from different initial conditions.

*Graduate Student, Faculty of Aerospace Engineering, Technion - Israel Institute of Technology, 3200003 Haifa, Israel maxf1000@gmail.com
†2Professor, Faculty of Aerospace Engineering, Technion - Israel Institute of Technology, 3200003 Haifa, Israel tal.shima@technion.ac.il

Reachability-Based Delayed Decision Guidance for Conventional Interceptors

Vitaly Shalumov*† Gleb Merkulov†‡ and Tal Shima§

Technion – Israel Institute of Technology, Haifa, 3200003, Israel

In engagements with multiple interceptors and multiple targets, in which the number of interceptors is larger than the number of targets, the excess of interceptors has to be managed. A beneficial strategy could be to divide the interceptor group into the pursuit and backup groups. The pursuit group covers all the incoming evaders, and the backup interceptors act upon the success of the pursuit group. This concept for a single trailing pursuer was introduced as delayed-decision guidance (DDG) by Turetsky et al. in [1], in which the trailing pursuer was guided in the first stage against all evader simultaneously to minimize the average control effort over all engagement decisions.

The obtained DDG guidance law does not address several important considerations. First, the structure of the guidance law is rather complex and requires re-configuration for each specific scenario. In addition, this approach relied on linearization around common initial line-of-sight; however, the realistic allocation would take place during the midcourse phase [2], in which the nonlinear kinematic model has to be considered. Moreover, the aforementioned DDG approach does not account for a specific guidance law used by the legacy interceptor and does not consider miss distance as a relevant figure of merit.

Therefore, this paper pursues the following goals. First, we generalize the DDG performance index to include the miss distance. Second, we extend the DDG application to concentrate on the specific aspects of the midcourse and terminal guidance. In the midcourse phase before the decision, our focus is on considering the nonlinearity of kinematics, while we disregard the dynamic response. In the terminal phase after the allocation decision, we adopt the standard linearization assumptions and account for the interceptor dynamics. The two phases are connected via the handover point specified by the position and heading, which is chosen in the proposed optimization procedure from the reachable set of the pursuer. The performance evaluation of decision point performance is based on adjoint simulation tools which drastically reduce the number of required simulations. Specifically, we demonstrate the optimization methodology for the case of the minimum-time pursuer trajectories in the midcourse phase and proportional navigation (PN) guidance law in the terminal phase.

For example, consider a two-vs-two engagement scenario in Fig. 1. The pursuer P_1 engages evader E_1, whereas the allocation of P_2 will be decided based on the intercept outcome between P_1 and E_1. Thus, during the midcourse phase it is guided to a decision point represented by the position and heading chosen from the reachable set, from which either of evaders can be engaged with favorable initial conditions.

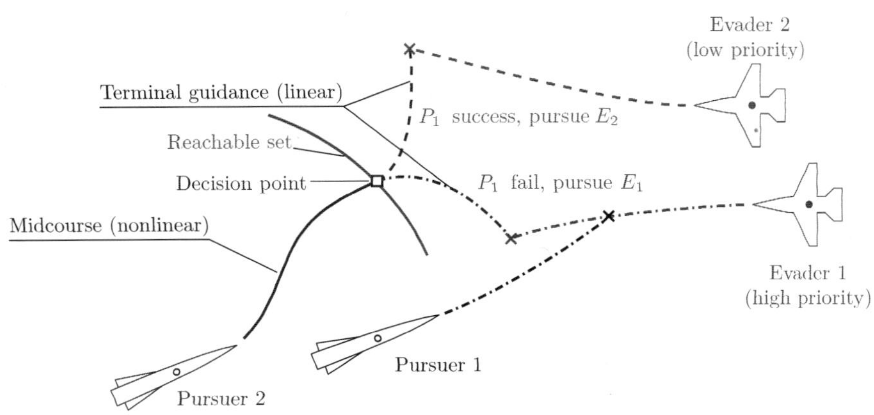

Fig. 1 Reachability-based Delayed Decision Guidance.

*Equal contribution.

†Doctoral Student, The Technion Program for Autonomous Systems and Robotics; vitaly.shalumov@gmail.com.

‡Doctoral Student, Faculty of Aerospace Engineering, glebmerkulov@campus.technion.ac.il.

§Professor, Faculty of Aerospace Engineering; tal.shima@technion.ac.il.

The simulation results for an example scenario are presented in the Fig. 2. Due to the high nonlinearity of the midcourse phase, significant trajectory shaping is done both for small and large α and the two trajectories differ significantly to account for miss-effort trade-off. The simulation results demonstrate a general tendency to prolong the terminal phase since it leads to smaller miss distance and control values.

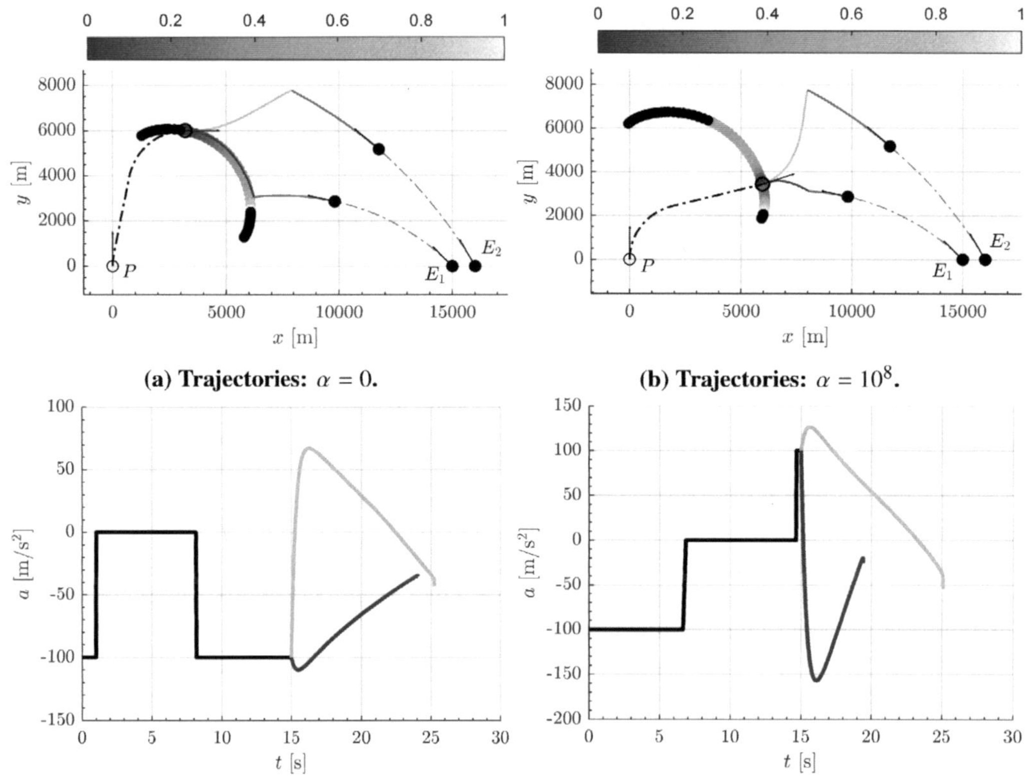

(a) **Trajectories: $\alpha = 0$.**

(b) **Trajectories: $\alpha = 10^8$.**

(c) **Acceleration: $\alpha = 0$.**

(d) **Acceleration: $\alpha = 10^8$.**

Fig. 2 Simulation results for tested Scenario. Colorbar depicts normalized cost value, which is a weighted sum of expected miss distance and control effort (α is the weight on miss distance). Blue and green lines represent trajectories and maneuvers against E_1 and E_2, respectively after the allocation.

References

[1] Turetsky, V., Weiss, M., and Shima, T., "Minimum Effort Pursuit Guidance with Delayed Engagement Decision," *Journal of Guidance, Control, and Dynamics*, Vol. 42, No. 12, 2019, pp. 2664–2670. https://doi.org/10.2514/1.G004393.

[2] Imado, F., Kuroda, T., and Miwa, S., "Optimal midcourse guidance for medium-range air-to-air missiles," *Journal of Guidance, Control, and Dynamics*, Vol. 13, No. 4, 1990, pp. 603–608. https://doi.org/10.2514/3.25376.

Guidance on Stationary Target with Range-Rate Information Only

Liat Peled-Eitan and Ilan Rusnak

RAFAEL (88), P.O.Box 2250, Haifa, Israel, 3102102.

Common target tracking and guidance use the angle measurement to steer the interceptor toward the target. These methods are well known and documented in the literature.

Here a method that uses only the range rate for target tracking and guidance applicable to an electromagnetic sensor is presented and analyzed. The method is based on the fact that the range rate is at maximum when the interceptor's velocity is aligned with the line-of-sight. This method is presented in [2,3] for tracking. Here the method is applied to guidance. This method realizes a Pure Pursuit guidance law [1] and enables the guidance of the interceptor from all aspects.

The idea is as follows: Assume a moving interceptor with a sensor measures the range rate to a stationary target. It is well known that when the interceptor moves along the line-of-sight (LOS) the range rate is at maximum. Thus if the interceptor velocity is kept aligned with the LOS, the target is tracked, the interceptor is guided toward the target, and the scenario ends in intercept. Thus only the range rate is used and angular measurements are not required.

The range rate enables computation of the deviation of the interceptor velocity from the LOS direction. However, this information does not give the sign of the deviation. Thus a respective maximum-seeking algorithm is developed.

The Pure-Pursuit guidance is implemented with a stiff neck sensor.

The performance is demonstrated by simulations.

The scenario is presented in Figure 1. The interceptor, M, moves with velocity, V_M, toward a stationary target, T. The line-of-sight angle (LOS) relative to a reference direction is, λ. The target-to-interceptor range is r_{TM}. The interceptor trajectory angle is γ_M and the angle between the LOS and the interceptor's velocity is σ_M. α is the angle of attack.

The equations of motion are

$$\sigma_M = \gamma_M - \lambda$$

$$\dot{r}_{TM} = -V_M \cos(\sigma_M)$$

$$r_{TM}\dot{\lambda} = -V_M \sin(\sigma_M)$$

$$\dot{\gamma}_M = \frac{a_M}{V_M}$$

$$\dot{\sigma}_M = \frac{V_M}{r}\sin(\sigma_M) + \frac{a_M}{V_M}$$

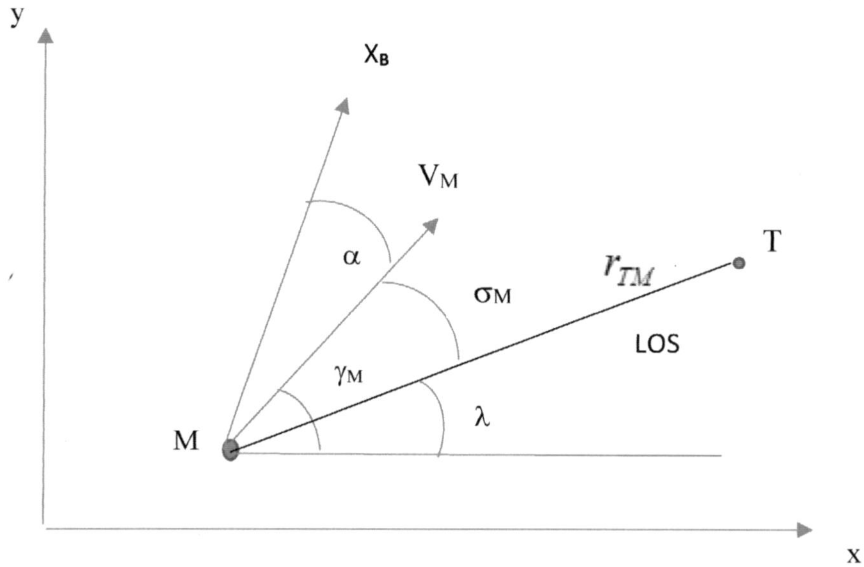

Figure 1: The kinematics for a stationary target.

The commanded acceleration is as follows:

From the measured range rate, $\tilde{\dot{r}}_{TM}$,

$$\tilde{\sigma}_M = \begin{cases} \left|\arccos\left(\frac{\tilde{\dot{r}}_{TM}}{V_M}\right)\right| & , \frac{\tilde{\dot{r}}_{TM}}{V_M} \leq 1 \\ 1 & , \frac{\tilde{\dot{r}}_{TM}}{V_M} > 1 \end{cases}.$$

Evaluate

$$s(t) = \begin{cases} s(t-h), & \tilde{\dot{r}}_{TM}(t) - \tilde{\dot{r}}_{TM}(t-h) > 0 \\ -s(t-h), & \tilde{\dot{r}}_{TM}(t) - \tilde{\dot{r}}_{TM}(t-h) < 0 \end{cases}.$$

where $h > 0$ is the sampling time.

Then the commanded acceleration is

$$a_{Mc}(t) = -s(t)k'(t)\sin(\tilde{\sigma}_M)$$

Figure 2 presents the trajectories of a missile guided toward a stationary target. The missile is launched from the origin at large initial aspect angles.

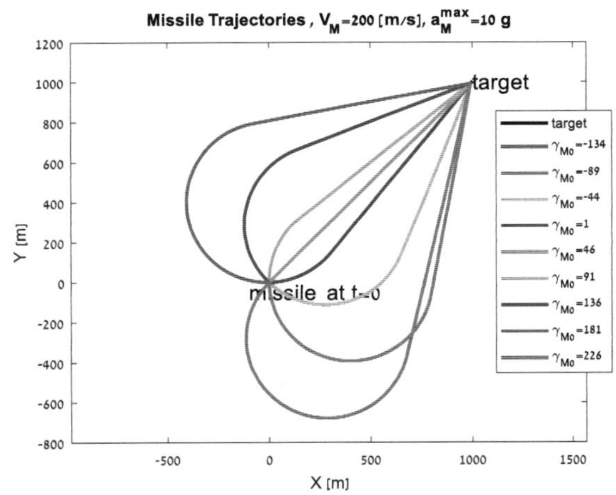

Figure 2: Interceptor trajectories for intercepting stationary target for different initial interceptor trajectory angles and ideal interceptor autopilot.

1] Shneydor, N.A.: Missile Guidance and Pursuit: Kinematics, Dynamics and Control, Horwood Publishing Limited, 1998.

2] J. Vaganay, P. Baccou and B. Jouvencel, "Homing by acoustic ranging to a single beacon," OCEANS 2000 MTS/IEEE Conference and Exhibition. Conference Proceedings (Cat. No.00CH37158), 2000, pp. 1457-1462 vol.2, doi: 10.1109/OCEANS.2000.881809.

3] Dave W. Oyler, Pierre T. Kabamba and Anouck R. Girard: Binary Range-Rate Measurements and Homing Guidance, Journal of Dynamic Systems, Measurement, and Control, ASME, APRIL 2015, Vol. 137.

Bearing-only Formation Control with Directed Sensing

Jiacheng Shi and Daniel Zelazo
Technion-Israel Institute of Technology

1 Abstract

This work investigates the formation control problem, which aims to drive a multi-agent system to achieve a desired spatial configuration. A control scheme called bearing-only formation control is proposed to solve this problem. This controller has attracted research interest because it requires and applies only bearing measurements between agents, rather than using distance measurements.

One limitation of bearing-only formation control is the sensing condition assumed. Early works extensively studied this controller assuming undirected sensing between agents, which is regarded as an unrealistic condition. More recent efforts have focused on the problem with directed sensing, where agent i can sense agent j but not necessarily vice versa. Recent research showed that the bearing-only formation control law also works with a specific class of directed sensing graphs called leader-first-follower (LFF) graphs generated from a bearing-based Henneberg construction. This was a major step forward from the previous assumption of undirected graphs to a limited class of directed graphs. Building on this, the goal of this work is to extend the results to more general conditions on the directed sensing graphs for which bearing-only formation control succeeds.

To achieve this, we first analyze a multi-agent system with one movable agent controlled by the bearing-only formation controller. The goal is to determine whether this movable agent will converge to a desired position satisfying all specified target bearings, for different numbers of outgoing edges (agents it can sense). The main challenge is finding equilibrium points of this nonlinear system. Our approach focuses on analyzing a related linear problem and establishing connections to equilibria of the nonlinear system. After determining the system equilibria, we also provide stability analysis showing that the movable agent converges to the desired position. This result then allows us to extend the class of graphs that solve the problem to LFF graphs where the follower agents can have more than 2 outgoing sensing edges. We provide stability and convergence analysis for this case. We also explore a further expansion of the class of graphs and conjecture that a sufficient condition for solving the more general formation control problem is that there must exist a subgraph that is a LFF graph. This conjecture is verified with simulation studies.

This work is concerned with the development and analysis of a very fast quaternion estimator from two vector observations. Processing as many observations as available may sound more efficient, but the insight from solutions with two observations is lost, and a marginal increase in accuracy might not pay off compared to the required computation burden. Besides, there is a growing class of very small aerospace systems, e.g. satellites or unmanned aerial vehicles, that accommodate two sensors only onboard, e.g. a magnetometer and a Sun sensor or a three-axis accelerometer. Previous works on attitude matrix determination from two vector observations include the TRIAD estimator, its covariance analysis, an optimized TRIAD estimator, and an optimal attitude matrix estimator. The EULER 2 estimator utilizes the Rodrigues formula to reconstruct the Euler axis/angle parameters and the attitude matrix. It invokes optimization of Wahba's loss function in the case of noisy measurements and exploits a coplanarity condition of the measured and predicted vectors to develop analytical formulas. These estimators focused on the attitude matrix, from which a quaternion estimate can be extracted if required. On the other hand, other estimators directly targeted the quaternion. A closed-form expression of the QUEST algorithm is developed for the case of two observations. Reynolds introduces a quaternion parameterization given a single vector observation and several methods are devised for analytical quaternion determination given two observations. Interestingly, the quaternions are represented as classes of equivalence where the elements are collinear but not necessarily unit-norm. Motivated by these insights, Markley devised very efficient algorithms: an optimal one, more accurate and as fast as TRIAD, and a suboptimal one, faster and as accurate as TRIAD. The QUEST algorithm, however, requires solving for the maximum eigenvalue in a characteristic equation, which is known to be less robust than more direct approaches in computing the associated eigenvector, and the related computation burden is greater than other similar quaternion estimators. The EULER2 estimator involves trigonometry, adding a burden to the computations, and it lacks singular cases mitigation. The first method of Reynolds seeks a quaternion in the span of two base quaternions, but no estimator is actually devised. While the second method exploits geometrical insights for developing a quaternion estimate, it breaks down for specific unaddressed singular cases. Markley's work addresses the singularities and its analytical formulas for the quaternion help with the computation load, but the optimization still adds to the calculations. More importantly, none of the above-mentioned works feature some error analysis, in particular for the bias and the covariance of the estimation error. In this work, we introduce a simple quaternion estimator that is faster and mostly as accurate as Markley's suboptimal estimator, along with a thorough error analysis, and an analytical solution of the singular cases. The quaternion is sought as the unique solution to a set of orthogonality conditions that involve simple nonlinear expressions of the measurements. This approach fundamentally differs from previous works because the quaternion is not sought in the span of a particular basis, and because the orthogonality conditions do not require solving for the optimal Wahba's loss value. The proposed algorithm is exceedingly simple, which dramatically simplifies the computations. This also enables systematic error analysis and the mitigation of singularity cases. In addition, this study sheds light on various hidden properties in quaternion estimation, like quaternion normalization, independently of the estimator studied here. The analysis is carried out in four dimensions both for multiplicative and additive errors and lends itself to second-order and fourth-order expressions for the biases and the covariance matrices, respectively, in terms of the measurement noises. The consistency of the approximations is verified via extensive Monte Carlo simulations for representative test cases. Additional Monte Carlo simulations evaluate the performance of the proposed estimator in terms of accuracy and computation burden. The novel estimator outperforms all reviewed quaternion estimators in terms of computations, and its accuracy may be better depending on the required reliability range.

Adjoint Differentiation Method for Trajectory Simulations with Terminal Conditions

Amir Mittelman[1], Ingo Jahn[2], Rowan Gollan[1]

Pushing the edge of aerospace flight vehicle capabilities requires the inclusion of multiple disciplinary and interdisciplinary considerations into the design process. In many cases, vehicle design optimization is performed for a single design point (flight conditions). Including trajectory simulation in the vehicle design optimization process effectively transforms it into a multi-point optimization providing mission-level performance objectives to be included in the decision-making process. Efficiently solving large-scale (large number of design variables) Multi-Disciplinary Analysis and Optimization (MDAO) problems is therefore sought after. Gradient-based optimization algorithms reduce the number of design iterations required for convergence by exploiting the objective function sensitivities (gradient) as well as the objective function value. The Adjoint method of differentiation is an extremely efficient method for calculating the gradients of large-scale time-dependent ODEs (dynamic systems).

In this work, we applied the well-established Lagrange-multiplier adjoint differentiation method of dynamic systems over constant time intervals to a flight vehicle simulation. In aerospace mission design, the flight duration is often dependent on the trajectory state variables (e.g. reaching the desired destination). Therefore, special consideration was added to the method to account for the commonly used terminal conditions on the state variables. A proof-of-concept demonstration of the extended method highlights its effective implementation as the partial derivatives of the trajectory simulation can be easily calculated analytically. The reduction in objective function evaluations needed for convergence provides validation of the efficiency of the optimization algorithm as well as the differentiation method used.

[1] School of Mechanical and Mining Engineering, Centre for Hypersonics, University of Queensland, St. Lucia QLD 4072, Australia

[2] Institute for Advanced Engineering and Space Sciences, Springfield Campus, The University of Southern Queensland, Toowoomba, QLD, 4350, Australia

Altruistic Kalman Filtering

Yuval Aldema-Tshuva, Ronny Shapiro, Nitai Stein And Yaakov Oshman

This work builds upon and extends the results of "Double-Opportunity Estimation via Altruism" (Stein and Oshman, 2022), where the authors only addressed the static case of estimating a random variable. Addressing the time-varying, state estimation problem, we present herein an altruistic Kalman filter for a stochastic process.

(Stein and Oshman, 2022) is a fundamental study that redefines the estimation problem and cost function, thus generalizing the common definition of minimum mean squared error estimation. The problem considers two information-sharing agents and proves that by altruistically cooperating, these agents can achieve a better overall mission performance, as compared with the case where both agents work separately (i.e., each optimizing its own, egoistic, performance). The altruistic cost function is defined as the expectation of the minimum between the two estimators' squared errors. Fundamental equations are derived for two types of altruistic cooperative estimation problems, corresponding to heterarchical (where both agents, with no preference, minimize a global cost function) and hierarchical (where one agent minimizes the regular, mean squared error (MSE) cost, and the other minimizes the global altruistic cost) setups. Although these equations are, generally, hard to solve, they can be readily solved in the Gaussian case, where the solution only requires the computation of the largest eigenvalue of the conditional covariance matrix and its corresponding eigenvector. The best overall estimation performance is achieved in the heterarchical setup. However, even the inferior hierarchical setup outperforms the non-cooperating setup, where each agent optimizes its own, egoistic, performance.

Addressing the dynamic, state estimation problem, in the proposed paper we present a cooperation-based, altruistic, Kalman filter. Both heterarchical and hierarchical setups are considered. A dynamic simulation study is conducted, addressing a ballistic missile interception scenario. The study demonstrates that the performance order governing the static case is valid in the dynamic case as well. Considering the maximal regular (egoistic) MSE as the baseline, the study shows that the corresponding altruistic performance measures are reduced by 18% and 30%, for the hierarchical and heterarchical configurations, respectively.

Robustness of the SDC based Quadratic Optimal Control of Nonlinear Systems

Ilan Rusnak and Maital Levy
RAFAEL , P.O.Box 2250, Haifa, Israel, 3102102.

The quadratic optimal control of a linear system had been solved more than 50 years ago. Recently an explicit closed-loop solution in finite time interval to the problem of optimal control of affine nonlinear systems for a quadratic criterion has been presented in [1,2]. This solution uses the State-Dependent Coefficient (SDC) form representation of affine nonlinear systems.

The solution is composed of two steps. The first step is solving the Two-Point Boundary Value problem (TPBVP) and getting the optimal state and the input. This optimal input is the open loop control. The second step for a closed loop solution requires a backward integration of a nonsymmetric nonlinear Riccati equation that derives the feedback gain matrix, exactly as for linear systems.

The objective of this paper is an analysis and a demonstration of the stability of the closed-loop control.

It is shown that the closed-loop nonlinear control in finite time possesses similar stability and robustness properties as the closed-loop control of linear systems.

The nth order affine nonlinear dynamic system dealt with is

$$\dot{x}(t) = f(x(t)) + Gu(t);\ x(t_o) = x_o, \qquad (1)$$

The problem being considered here is the optimal control of the system (1). That is we wish to minimize the quadratic criterion

$$J(t) = \frac{1}{2}\left\{ x(t_f)^T P_f x(t_f) + \int_t^{t_f} \left[x(\tau)^T Qx(\tau) + u(\tau)^T Ru(\tau) \right] d\tau \right\}$$

with respect to, *u(t)*, subject to (1).

The system (1) can be represented by the use of the State-Dependent Coefficient (SDC) form, which is defined (when $f(0)=0$,) as

$$f\left(x(t)\right) =: F\left(x(t)\right)x(t)$$

Then, the solution is given by the Two Point Boundary Value Problem (TPBVP)

$$\dot{x}(t) = F\left(x(t)\right)x(t) - GR^{-1}G^T\lambda(t); \qquad x(t_o) = x_o$$
$$\dot{\lambda}(t) = -Qx(t) - f_x\left(x(t)\right)^T \lambda(t); \qquad \lambda(t_f) = P_f\ x(t_f),$$

and the optimal full-state feedback input is given by

$$u^*(t) = -R^{-1}G^T P^*(t) x^*(t)$$

where *P(t)* is given by the nonsymmetric Riccati equation that is integrated backward in time

$$-\dot{P} = PF\left(x^*(t)\right) + f_x\left(x^*(t)\right)^T P + Q - PGR^{-1}G^T P, \quad P(t_f) = P_f.$$

For stability and robustness, we apply the input

$$\tilde{u}(t) = -\gamma R^{-1} G^T P(x^*(t))\tilde{x}(t),$$

and use the Lyapunov function

$$V = \frac{1}{2}\tilde{x}^T \left(P^* + P^{*T}\right)\tilde{x}.$$

Then

$$\dot{V} = \frac{1}{2}\tilde{x}^T \left\{ \begin{array}{l} -2\left(Q + [2\gamma - 1] P_v^T G R^{-1} G^T P_v\right) \\ +\left(F_v(\tilde{x}) - F_v(x^*)\right)^T \left(P_v + P_v^T\right) + \left(P_v + P_v^T\right)\left(F_v(\tilde{x}) - F_v(x^*)\right) \\ +\left(F_v(x^*) - f_x(x^*)\right)^T P_v + P_v^T \left(F_v(x^*) - f_x(x^*)\right) \end{array} \right\} \tilde{x}$$

This means that:

(i) as long as the difference between the Jacobian and the SDC form, $F(x) - f_x(x)$, is not too big;

(2) the optimal trajectory and the actual trajectory, $F_v(\tilde{x}) - F_v(x^*)$, are close;

(3) $\gamma > 1/2$; and

(4) Q is sufficiently big;

the stability/robustness of the closed loop is guaranteed.

For demonstration, the 2nd order nonlinear chaotic Van der Pol equation is used. Figure 1 presents the change in the plant's states due to the plant's (c) parameter deviation from its nominal value. Figure 2 demonstrates the non-symmetry of the feedback gain matrix. The closed loop remains stable and the terminal conditions are quite closely satisfied. In the paper, additional sensitivity tests are presented.

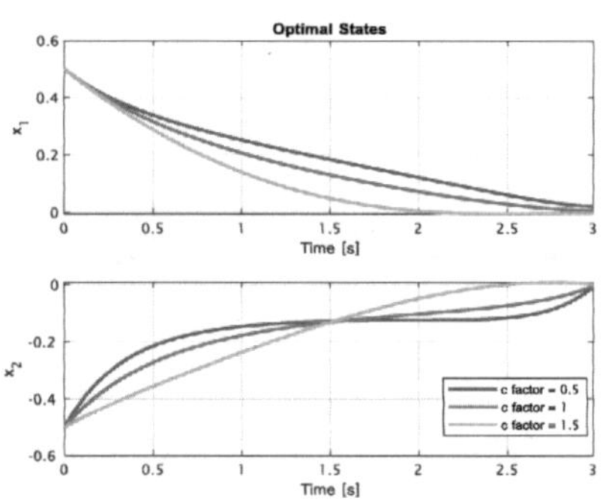

Figure 1: The states of the plant due to ±50% change of the (c) parameter.

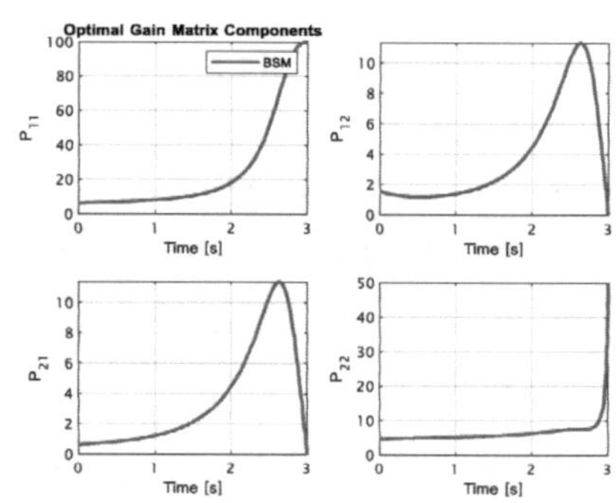

Figure 2: The asymmetric feedback gain matrix, *P(t)*.

References

1) Ilan Rusnak and Maital Levy: A Formal Solution of the Quadratic Optimal Control for Nonlinear Systems, 29th Mediterranean Conference on Control and Automation (MED 2021) 1, June 22 - 25 2021, Bari, Puglia, Italy. DOI: 10.1109/MED51440.2021.9480204

2) Ilan Rusnak and Maital Levy: State-Dependent Coefficient form Based Solution of the Quadratic Optimal Control for Nonlinear Systems, the 61st Israel Annual Conference on Aerospace Science, March 9-10, 2022, Israel.

Diodes and the Importance of Network Orientations in Diffusively-Coupled Networks

Feng-Yu Yue*[1] and Daniel Zelazo†[1]

[1]Faculty of Aerospace Engineering, Technion

Abstract

Diffusively-coupled networks, a common architecture for multi-agent systems, are used in many foundational cooperative control problems, such as consensus and formation control. One reason for its widespread use is that this structure contains information on agents, edge controllers, and the latent interaction network between agents, making passivity theory a natural tool for its analysis. However, an implicit property of this structure is the symmetric feedback interconnection of the incidence matrix, which erases any orientation for the network that may be encoded in the incidence matrix. So, the diffusively-coupled structure is useful when modeling undirected networks, and the existing passivity-based analytical framework is primarily effective for undirected networks. These present the challenge of how to model directed networks effectively as an unresolved issue.

In this work, inspired by the single conductance property of a diode, we explore a special instance of diffusive networks where the edge controllers follow the model of an ideal diode. The diffusive diode networks highlight the importance of the chosen edge orientation with respect to the incidence state values and exhibit system behaviors that correspond to directed networks or undirected networks. To enable a more direct comparison to the undirected and directed consensus protocols, we focus on the case where the agents are modeled by integrator dynamics.

We begin with a basic analysis of the networked system with diodes. We show that the generated trajectories are bounded and continuous, and the closed-loop system can be expressed as a linear consensus protocol over a time-varying graph. This is followed by an example illustrating the complexity of diffusive diode networks. With these observations, we present a sufficient condition on when the diffusive diode networks can achieve average consensus, which does not highlight the role of the initial conditions or the orientation of the graphs. Then, for some special directed graphs (i.e., paths, cycles, rooted out-trees), we discuss how the system state together with the graph orientation, influences the network steady-state. We explore sufficient conditions on the graph and the initial conditions of the network that lead to consensus. We also provide a negative result providing a necessary condition that graphs containing rooted out-trees can not achieve consensus. Our results are supported by numerous numerical examples.

*fengyu.yue@campus.technion.ac.il. (Corresponding author: Feng-Yu Yue)

†dzelazo@technion.ac.il

Application of the Riccati Inequality to design a controller for stabilization of nonlinear affine system

György Hexner and Ilan Rusnak, RAFAEL.

May 3, 2024

1 Extended Abstract

Let the system to be controlled be

$$\dot{\boldsymbol{x}} = \boldsymbol{f}(\boldsymbol{x}) + \boldsymbol{B}(\boldsymbol{x})\boldsymbol{u}, \boldsymbol{x}(0) = \boldsymbol{x_0} \tag{1.1}$$

with $\boldsymbol{f}(\boldsymbol{x})$, and $\boldsymbol{B}(\boldsymbol{x})$ matrices with polynomial entries. The system is controllable.

The system (1.1) can be recast into the State Dependent Coefficient (SDC) form

$$\dot{\boldsymbol{x}} = \boldsymbol{F}(\boldsymbol{x})\boldsymbol{x} + \boldsymbol{B}(\boldsymbol{x})\boldsymbol{u} \tag{1.2}$$

The SDC form of (1.1) is not unique [3] , so the following depends on the specific selection of the SDC form.

We wish to find a control such that the criterion

$$J = \frac{1}{2}\int_0^{\infty} \left(\boldsymbol{x}^T\boldsymbol{Q}\boldsymbol{x} + \boldsymbol{u}^T\boldsymbol{R}\boldsymbol{u}\right) dt \tag{1.3}$$

with $\boldsymbol{Q} \succ \boldsymbol{0}$ and $\boldsymbol{R} \succ \boldsymbol{0}$ is minimized for the Lyapunov function

$$V = \frac{1}{2}\boldsymbol{x}^T\boldsymbol{P}\boldsymbol{x} \tag{1.4}$$

where $\boldsymbol{P}$ is a **constant** positive definite matrix.

The paper proposes a novel approach for derivation of a control law that leads to global asymptotic stability. This approach is based on use of SDC, Hamilton-Jacobi inequality (HJI), Riccati inequality (RI) and the lately introduced optimization tools based on representation of nonlinear systems by Sum-Of-Squares (SOS).

To find the optimal control law the HJB equation

$$-\frac{\partial V}{\partial t} = \min_u \left\{\frac{1}{2}\boldsymbol{x}^T\boldsymbol{Q}\boldsymbol{x} + \frac{1}{2}\boldsymbol{u}^T\boldsymbol{R}\boldsymbol{u} + \frac{\partial \boldsymbol{V}}{\partial \boldsymbol{x}}\left[\boldsymbol{F}(\boldsymbol{x})\boldsymbol{x} + \boldsymbol{B}(\boldsymbol{x})\boldsymbol{u}\right]\right\} \tag{1.5}$$

should be solved. For (1.4) this gives

$$\min_u \left\{ \frac{1}{2} \boldsymbol{x}^T \boldsymbol{Q} \boldsymbol{x} + \frac{1}{2} \boldsymbol{u}^T \boldsymbol{R} \boldsymbol{u} + \frac{\partial \boldsymbol{V}}{\partial \boldsymbol{x}} \left[\boldsymbol{F}(\boldsymbol{x}) \boldsymbol{x} + \boldsymbol{B}(\boldsymbol{x}) \boldsymbol{u} \right] \right\} = 0 \tag{1.6}$$

that should be satisfied. Then the optimal input is

$$\boldsymbol{u} = -\boldsymbol{R}^{-1} \boldsymbol{B}(\boldsymbol{x})^T \left[\frac{\partial \boldsymbol{V}}{\partial \boldsymbol{x}} \right]^T . \tag{1.7}$$

and from (1.4) we have

$$\frac{\partial \boldsymbol{V}}{\partial \boldsymbol{x}} = \boldsymbol{x}^T \boldsymbol{P}. \tag{1.8}$$

substituting (1.7) and (1.8) into (1.6) we get that $\boldsymbol{P}$ has to satisfy the Riccati Equality (RE)

$$\boldsymbol{P} \boldsymbol{F}(\boldsymbol{x}) + \boldsymbol{F}(\boldsymbol{x})^T \boldsymbol{P} + \boldsymbol{Q} - \boldsymbol{P} \boldsymbol{B}(\boldsymbol{x}) \boldsymbol{R}^{-1} \boldsymbol{B}(\boldsymbol{x})^T \boldsymbol{P} = \boldsymbol{0} \tag{1.9}$$

This RE usually does not have a solution for a constant $\boldsymbol{P}$.
Thus, we propose to find a constant $\boldsymbol{P}$ that satisfies the following HJI

$$\min_u \left\{ \frac{1}{2} \boldsymbol{x}^T \boldsymbol{Q} \boldsymbol{x} + \frac{1}{2} \boldsymbol{u}^T \boldsymbol{R}^{-1} \boldsymbol{u} + \frac{\partial \boldsymbol{V}}{\partial \boldsymbol{x}} \left[\boldsymbol{F}(\boldsymbol{x}) \boldsymbol{x} + \boldsymbol{B}(\boldsymbol{x}) \boldsymbol{u} \right] \right\} \leq 0 \tag{1.10}$$

that leads to the RI

$$\boldsymbol{P} \boldsymbol{A}(\boldsymbol{x}) + \boldsymbol{A}^T(\boldsymbol{x}) \boldsymbol{P} - \boldsymbol{P} \boldsymbol{B}(\boldsymbol{x}) \boldsymbol{R}^{-1} \boldsymbol{B}^T(\boldsymbol{x}) \boldsymbol{P} + \boldsymbol{Q} \preceq \boldsymbol{0} \; \forall \boldsymbol{x}. \tag{1.11}$$

If a solution of (1.11) exists then the control is

$$\boldsymbol{u} = -\boldsymbol{R}^{-1} \boldsymbol{B}^T(\boldsymbol{x}) \boldsymbol{P} \boldsymbol{x} \tag{1.12}$$

This control law stabilizes the system (1.1).
A simple example is as follows

$$\dot{x} = \sin(x) + u \tag{1.13}$$

A constant feedback

$$u = -k\,x, \quad k > 1 \tag{1.14}$$

leads to a global exponential stability.

To see that (1.12) leads to global stability, consider the candidate Lyapunov function (1.4). Since $\boldsymbol{P}$ is positive definite, all that remains to prove is that the derivative of the candidate Lyapunov function is negative definite. The derivative is

$$\dot{V} = \frac{1}{2} \boldsymbol{x}^T \boldsymbol{P} \dot{\boldsymbol{x}} + \frac{1}{2} \dot{\boldsymbol{x}}^T \boldsymbol{P} \boldsymbol{x} \tag{1.15}$$

then

$$\dot{\boldsymbol{x}} = \boldsymbol{A}(\boldsymbol{x})\boldsymbol{x} - \boldsymbol{B}(\boldsymbol{x})\boldsymbol{R}^{-1}\boldsymbol{B}(\boldsymbol{x})^T\boldsymbol{P}\boldsymbol{x} \tag{1.16}$$

and

$$\dot{V} = \frac{1}{2}\boldsymbol{x}^T\left[\boldsymbol{P}\boldsymbol{A}(\boldsymbol{x}) + \boldsymbol{A}(\boldsymbol{x})^T\boldsymbol{P} - \boldsymbol{P}\boldsymbol{B}(\boldsymbol{x})\boldsymbol{R}^{-1}\boldsymbol{B}(\boldsymbol{x})^T\boldsymbol{P} - \boldsymbol{P}\boldsymbol{B}(\boldsymbol{x})\boldsymbol{R}^{-1}\boldsymbol{B}(\boldsymbol{x})^T\boldsymbol{P}\right]\boldsymbol{x}. \tag{1.17}$$

Using (1.11) we get

$$\dot{V} \leq -\frac{1}{2}\boldsymbol{x}^T\left[\boldsymbol{Q} + \boldsymbol{P}\boldsymbol{B}(\boldsymbol{x})\boldsymbol{R}^{-1}\boldsymbol{B}(\boldsymbol{x})^T\boldsymbol{P}\right]\boldsymbol{x} \tag{1.18}$$

That means that (1.16)is stable as $\dot{V}$ is negative for $x \neq 0$.

It is detailed that a constant $\boldsymbol{P}$, that satisfies the RI (1.11), is derivable by the application of the SOS theory. A SOS polynomial, $p(x)$, is a polynomial of the form

$$p(x) = \sum_i q_i^2(x) \tag{1.19}$$

where q_i are polynomials.

Efficient algorithms had been developed for finding a constant $\boldsymbol{P}$ in SOSTOOLS [4], YALMIP [2], JULIA[1] and more, for systems that assume the structure of SOS. Specifically, for solvability the negative of the left hand side of (1.11) has to be SOS. This depends on the system (1.1) and on the specific selection of the SDC form $\boldsymbol{F}(\boldsymbol{x})$ in (1.2).

However, the set of solutions of the RI is infinite so additional constraint is needed. Recall that the HJI (1.10) at the optimum satisfies

$$\frac{1}{2}x^TQx + \frac{1}{2}u^TRu + \frac{\partial V}{\partial x}\left[F(x)x + B(x)u\right] \leq 0 \tag{1.20}$$

Now consider the value function

$$\begin{gathered} V(\boldsymbol{x}(T)) - V(\boldsymbol{x}(0)) = \textstyle\int_0^T \dot{V}dt \\ = \textstyle\int_0^T \frac{\partial V}{\partial \boldsymbol{x}}\left(\boldsymbol{F}(\boldsymbol{x})\boldsymbol{x} + B(\boldsymbol{x})\boldsymbol{u}\right)dt \leq -\frac{1}{2}\int_0^T\left(\boldsymbol{x}^TQ\boldsymbol{x} + \boldsymbol{u}^T\boldsymbol{R}\boldsymbol{u}\right)\,dt \end{gathered} \tag{1.21}$$

where (1.20) is used. (1.3). Rearranging,

$$V(\boldsymbol{x}(T)) - V(\boldsymbol{x}(0)) \leq -\frac{1}{2}\int_0^T\left(\boldsymbol{x}^TQ\boldsymbol{x} + \boldsymbol{u}^T\boldsymbol{R}\boldsymbol{u}\right)\,dt \tag{1.22}$$

$$\frac{1}{2}\int_0^T\left(\boldsymbol{x}^TQ\boldsymbol{x} + \boldsymbol{u}^T\boldsymbol{R}\boldsymbol{u}\right)\,dt \leq V(\boldsymbol{x}(0)) - V(\boldsymbol{x}(T)) \tag{1.23}$$

As the HJI (1.10) and RI (1.11) impose (at least) asymptotic stability then we have $\lim\limits_{T\to\infty} V(\boldsymbol{x}(T)) = 0$. Finally,

$$\frac{1}{2}\int_0^T \left(\boldsymbol{x}^T Q\boldsymbol{x} + \boldsymbol{u}^T \boldsymbol{R}\boldsymbol{u}\right)\, dt \leq \frac{1}{2}\boldsymbol{x}(\boldsymbol{0})^T \boldsymbol{P}\boldsymbol{x}(\boldsymbol{0}) \tag{1.24}$$

thus we would like to minimize the upper bound on the cost. Notice that $\boldsymbol{x}_0^T \boldsymbol{P}\boldsymbol{x}_0 = trace\left\{\left[\boldsymbol{x}_0\boldsymbol{x}_0^T\right]\boldsymbol{P}\right\}$, thus we impose as a constraint the requirement that for some $\boldsymbol{T} \succ \boldsymbol{0}$

$$\min_{\boldsymbol{P}} trace(\boldsymbol{T}\boldsymbol{P}) \tag{1.25}$$

will be minimal (trace is a norm for positive matrices). The matrix $\boldsymbol{T} \succ \boldsymbol{0}$ reflects some preference on the convergence of the states.

To summarize, we wish to minimize 1.25 subject to

1. $\boldsymbol{P} \succ \boldsymbol{0}$;

2. $\boldsymbol{P}\boldsymbol{A}(\boldsymbol{x}) + \boldsymbol{A}^T(\boldsymbol{x})\boldsymbol{P} - \boldsymbol{P}\boldsymbol{B}(\boldsymbol{x})\boldsymbol{R}^{-1}\boldsymbol{B}^T(\boldsymbol{x})\boldsymbol{P} + \boldsymbol{Q} \preceq \boldsymbol{0}$.

In the presentation the JULIA [1] software is applied to derive a full state feedback with constant $\boldsymbol{P}$ that stabilizes the Van der Pol equation in an elliptical level set of Lyapunov function, that is included in $-3 \leqslant x_1 \leqslant 3, -4 \leqslant x_2 \leqslant 4$.

References

[1] Jeff Bezanson, Alan Edelman, Stefan Karpinski, and Viral B Shah. Julia: A fresh approach to numerical computing. *SIAM Review*, 59(1):65–98, 2017.

[2] J. Lofberg. YALMIP: A toolbox for modelling and optimization in MATLAB. In *Proceedings of 2004 IEEE International Symposium on Computer Aided Control System Design*, pages 284–289, 2004.

[3] Liat Peled-Eitan and Ilan Rusnak. Issues in design and selection of state dependent coefficient form for efficient estimation of nonlinear systems. 59th Israel Annual Conference on Aerospace Sciences, IACAS 2019, pages 1310–1335, 2019. 6-7 March.

[4] S. Prajna, A. Papachristodoulou, P. Seiler, and P. A. Parrilo. *Positive Polynomials in Control*, chapter SOSTOOLS and its Control Applications, pages 273–292. Springer, 2005.

Simplified Continuous High Dimensional Belief Space Planning with Adaptive Probabilistic Belief-dependent Constraints

Andrey Zhitnikov[1] and Vadim Indelman[2]

[1]Technion Autonomous Systems Program (TASP) [2] Department of Aerospace Engineering

Technion - Israel Institute of Technology, Haifa 32000, Israel

andreyz@campus.technion.ac.il , vadim.indelman@technion.ac.il

Online decision-making under uncertainty in partially observable domains, also known as Belief Space Planning, is a fundamental problem in Robotics and Artificial Intelligence. Due to an abundance of plausible future unravelings, calculating an optimal course of action inflicts an enormous computational burden on the agent. Moreover, in many scenarios, e.g., Information gathering, it is required to introduce a belief-dependent constraint. Prompted by this demand, in this paper, we consider a recently introduced probabilistic belief-dependent constrained POMDP. We present a technique to adaptively accept or discard a candidate action sequence with respect to a probabilistic belief-dependent constraint, before expanding a complete set of sampled future observations episodes and without any loss in accuracy (Figure 1). Moreover, using our proposed framework, we contribute an adaptive method to find a maximal feasible return (e.g., Information Gain) in terms of Value at Risk and a corresponding action sequence, given a set of candidate action sequences, with substantial acceleration. On top of that, we introduce an adaptive simplification technique for a probabilistically constrained setting. Such an approach provably returns an identical-quality solution while dramatically accelerating the online decision-making. Our universal framework applies to any belief-dependent constrained continuous POMDP with parametric beliefs, as well as nonparametric beliefs represented by particles. In the context of an information-theoretic constraint, our presented framework stochastically quantifies if a cumulative Information Gain along the planning horizon is sufficiently significant (e.g. for, Information gathering, active SLAM). As a case study, we apply our method to two challenging problems of high dimensional Belief Space Planning: active SLAM and Sensor Deployment. Extensive realistic simulations corroborate the superiority of our proposed ideas.

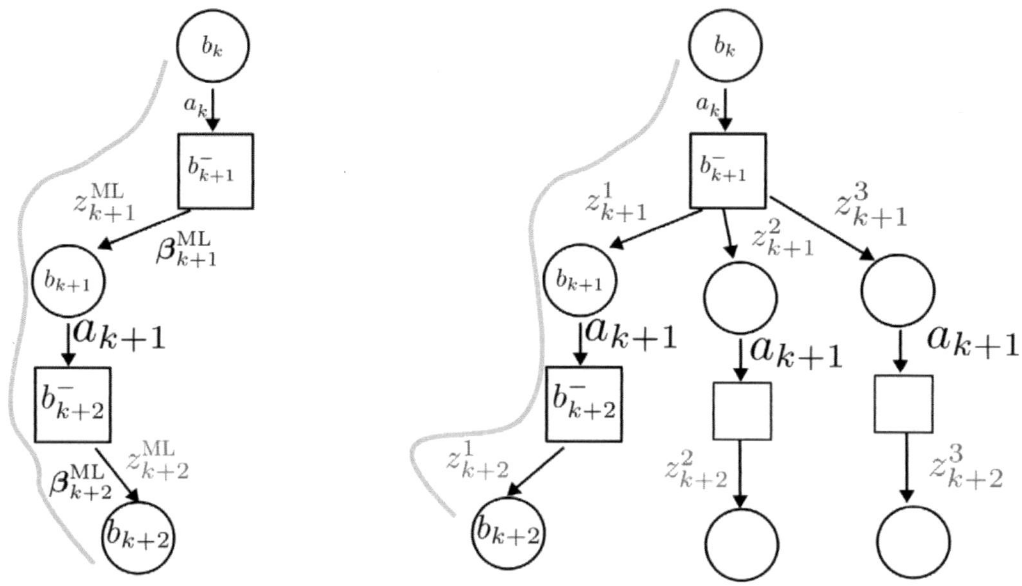

Figure 1. Illustration of the commonly used approach for information-theoretic constraint for stopping exploration versus ours. From the left, we visualize a belief tree approximated by the single maximum likely lace. In contrast, our approach approximates the belief tree by many laces.

References

[1] J. A. Placed, J. Strader, H. Carrillo, N. Atanasov, V. Indelman, L. Carlone, and J. A. Castellanos, "A survey on active simultaneous localization and mapping: State of the art and new frontiers," IEEE Trans. Robotics, vol. 39, no. 3, pp. 1686–1705, 2023.

[2] J. A. Placed and J. A. Castellanos, "Enough is enough: Towards autonomous uncertainty-driven stopping criteria," arXiv preprint arXiv:2204.10631, 2022.

Nonlinear Control of an Air-Breathing Hypersonic Vehicle Subject to Scramjet Engine Constrains

Abstract submitted to the IACAS 63

Ofir Vaknin* and Moshe Idan†
Faculty of Aerospace Engineering, Technion - Israel Institute of Technology
Haifa 3200003, Israel

April 2024

Hypersonic flight has gained significant attention in recent years due to its promising applications in the civil and military fields. Typically, such flight vehicles are powered by supersonic combustion ramjet (scramjet) engines. Control of these scramjet-powered, air-breathing hypersonic vehicles (AHVs) is a complex task due to the highly nonlinear nature of their dynamics and the inherent coupling between their aerodynamics and propulsion. Several attempts were made to address the longitudinal control of such systems, the dynamic and aerodynamic modeling of which is extremely challenging. Most of these studies focused on the tracking performance of the vehicle and did not sufficiently emphasize the coupling between the scramjet propulsion system and the AHV aerodynamics. The scramjet is very sensitive to the flow conditions around the inlet of the engine, in particular to changes in the angle of attack (AOA) of the vehicle, and can stop operating once it reaches Mach-number dependant limits. Thus, a suitable controller of the AHV should ensure good tracking as well as limit the system to a safe set dictated by the propulsion system characteristics.

This study presents a nonlinear stabilizing controller for the AHV that ensures its safety and proper flight characteristics dictated by the scramjet engine limitations. The controller was constructed using a well-known model for the longitudinal dynamics of the AHV with the addition of a canard and small modifications. A canard was added in previous works to counteract non-minimum phase dynamics and to increase control authority. In the current study, the use of the canard together with several mild model approximations yields a feedback linearizable form of the control-design-oriented vehicle dynamics. This facilitates the design of a simpler controller compared to the previously reported ones in the literature.

The control objectives are described by a control Lyapunov function (CLF) whereas the safety objective is specified by a control barrier function (CBF). Unifying these stability and safety objectives is executed through a quadratic program (QP). This QP is solved analytically to yield a controller that minimizes a pre-determined cost function and, thus, also exhibits a certain degree of optimality. The resulting constrained controller was validated in simulation and demonstrated good behavior in terms of stability and safety. Simulations of a similar controller with no imposed system limitations were also conducted. Naturally, the latter demonstrated safety violations of the

*M.Sc. Student. Email: ofir.v@campus.technion.ac.il
†Email moshe.idan@technion.ac.il

transient response of the system. The comparison of the two controllers shows that the constrained controller converges similarly to the unconstrained one, i.e., the scramjet engine limitations were addressed with only a slight compromise of the vehicle tracking performance. In conclusion, this study presents a relatively simple nonlinear analytic controller for the AHV that guarantees its safety with little compromise to tracking performance.

Model-Based Mesh Generation for Orthogonal Collocation Transcription

Ilan Taub, Vitaly Shaferman, Joseph Z. Ben-Asher

The proposed paper deals with a model-based approach for the efficient transcription of Orthogonal Collocation (Pseudospectral) methods in Optimal Control.

Orthogonal Collocation methods have been successfully used in the past two decades for the purpose of solving optimal control problems numerically. These methods transcribe a continuous time problem into a static parametric problem by approximating the problems' states using global polynomials. The resulting parametric optimization problem can then be solved using Nonlinear Programming algorithms (NLP). For these methods to work efficiently, the state approximating polynomials must be employed on smooth intervals. Since most realistic problems are only piecewise smooth, and the locations of discontinuities are not known in advance, current-day solutions employ mesh refinement algorithms that attempt to identify non-smoothness and divide the time domain into intervals accordingly. These algorithms then attempt to assess the minimal degree polynomial that approximates the states accurately enough at each interval. This results in a two-loop search, where the external loop refines the mesh, and the internal loop solves the NLP problem for each mesh iteration.

In general, non-smoothness occurs due to control switching and/or model-related discontinuities. Model-related discontinuities can occur due to a physical phenomenon, such as a sharp change in aerodynamic coefficients under transonic flow conditions, but they mostly occur due to the representation of the models. Many realistic problems have models that are based on empirical results, represented in the form of tabulated data, as they cannot be obtained analytically. Because the various computations of the parametric problem are most likely to be performed at intermediate points, the tabulated data is interpolated. The interpolation is usually carried out using piecewise continuous functions, as the data cannot be approximated accurately enough at the sampling points by a single approximating function. This means that some high derivative of these functions will be discontinuous at the tabulated data points and produce non smooth problems.

Current-day mesh refinement algorithms employ various discontinuity detection tests, which are often threshold-based. Some of these tests only identify the existence of a discontinuity but not its location. As a result, false detection or no detection can occur. The polynomial degree assessment in most of these algorithms is usually based on relative tests, which only allow incremental corrections. These deficiencies result in a mesh that continuously grows, creating large NLP problems and prolonging the overall convergence times.

In the proposed paper, we provide a novel model-based approach that identifies model-related discontinuities accurately based on a known model data representation. We also propose a novel polynomial degree assessment based on the high accuracy of the Gaussian Quadrature, which is an absolute assessment rather than incremental. This allows defining a mesh generation algorithm that provides well-fitted meshes, resulting in small NLP problems and reducing overall convergence time. The paper will demonstrate this approach on an optimization problem without control. In future work, control-related discontinuities will be treated as well in order to provide a complete solution.

The problem of planning for partially observable Markov decision processes (POMDPs) has garnered significant attention in recent years. The approaches employed in finding a solution to the POMDP vary depending on if the state space, observation space, and or action space are discrete, continuous, or a mixture. Additionally, there are often simplifying assumptions about the belief being Gaussian. With or without these assumptions, the problem of finding a globally optimal solution remains intractable for all but the smallest problems as, among other things, it requires reasoning about the random variables as defined by the POMDP. This poses great difficulties for low dimensional settings and is an even exponentially greater difficulty for the high-dimensional setting.

To find a tractable solution to the problem, tractable bounds on the reward or value function, with guarantees, are an attractive alternative to the explicit calculations required of the optimal solution. Such bounds find their origin in probability theory. The two common inequalities employed in the field of robotics and ai are Markov's inequality, which allows for bounding the expectation, and Hoeffding's inequality, which bounds the expectation with respect to an estimator with some probability.

In this work, under the belief that efficient planning originates from efficient probability theory bounds; we formulate and prove our own novel bounds in probability theory. We first formulate a bound between expectation and the partial expectation and discuss the computation complexity associated with the partial expectation. Leveraging these bounds, we go on to formulate novel probabilistic bounds that incorporate Hoeffding's inequality. These bounds allow the expectation to be bounded with respect to an estimator, not of the expectation, but rather either some conditional expectation or the partial expectation. We go on to provide conditions under which these bounds improve upon Hoeffding's inequality.

Having provided our general use novel bounds, we are ready to apply them in planning. We prove how a general value function can be bounded in a recursive manner via the use of partial expectation. After visiting this general scenario, we investigate the more specific scenario of information theoretical reward, which is known to be a more challenging problem than state dependent rewards. In this scenario we formulate bounds on the immediate reward that allows us to bound the value function. Finally, we look to address the SLAM (Simultaneous Localization and Mapping) problem, where reasoning about realization of data association is necessary and where data association induces a topology as can be seen in the respective factor graph. In this case we devise novel bounds on the value function that allow for reasoning over only part of the data association with guarantees. For convenience we summarize our main contributions.

- Formulate and prove novel theoretical bounds between expectation and partial expectation.
- Formulate and prove novel theoretical bounds between expectation and estimated partial expectation, with condition upon which they improve upon Hoeffding's inequality.
- Novel bounds on the conditional entropy derived from partial expectation.
- Exploiting similar topologies to allow for calculation reuse between rewards.

Dynamic Testing of Underwater Hybrid Ram Rockets

Sagi Dinisman*
Technion - Israel Institute of Technology, Haifa, 32000, Israel

Nachum E. Eisen†
IAI – Israel Aerospace Industry, Lod, 70100, Israel

Alon Gany‡
Technion - Israel Institute of Technology, Haifa, 32000, Israel

I. Introduction

Propelling high-speed submerged marine vehicles, such as torpedoes at very high velocities (of the order of 100 m/s), requires powerful propulsion systems to overcome the tremendous hydrodynamic drag. While rocket motors can provide adequate specific thrust, their energetic performance (specific impulse I_{sp}) and, hence, operating time and range, are relatively low. The marine ram rocket (ducted rocket), augmenting the rocket motor by ingesting water from the surrounding due to the vehicle motion, can substantially increase the specific thrust and specific impulse, hence, increasing the operating time and range. Its operating principle is similar to that of the aeronautical air-breathing ramjet, where both motors generate high thrust and energetic performance by accommodating the surrounding medium utilizing the high dynamic (ram) pressure resulting from the high velocity. Due to the high temperature in the combustion chamber, the incoming water boils, generating steam. This steam, along with the combustion products, is then accelerated through the nozzle, providing the necessary thrust for vehicle propulsion.

II. Experimental set-up and test procedure

For testing the water-breathing ramjet in actual cruise conditions, a unique water basin was used (as illustrated in figure 1). To guide the tested motor in a circular trajectory during the experiment and regulate its velocity, the submerged test motor is attached to a central horizontal rotating arm controlled by an electric motor.

The water-breathing ramjet used employed a hybrid propellant, utilizing polyester as the fuel and gaseous oxygen as the oxidizer. The motor inlet featured a hydrodynamically shaped polyester cover with a central hole to allow water passage through ¼" stainless steel tube. Oxygen from high-pressure cylinder located on the rotating boom was distributed into the central port of the solid fuel grain. At the rear end of the fuel, a stainless steel aft-mixing chamber was attached, enabling evaporation and effective mixing of incoming water with the combustion products before accelerating through a convergent choked copper nozzle with a throat diameter varying from 6.85 mm to 8 mm. Figure 2 illustrates a schematic of the motor.

* MSc student, Faculty of Aerospace Engineering, E-mail: sagidinisman@campus.technion.ac.il
† PhD, Aerospace Engineer, Missiles & Space Group, E-mail: nachum.eisen@gmail.com
‡ Professor Emeritus, Faculty of Aerospace Engineering, E-mail: gany@tx.technion.ac.il

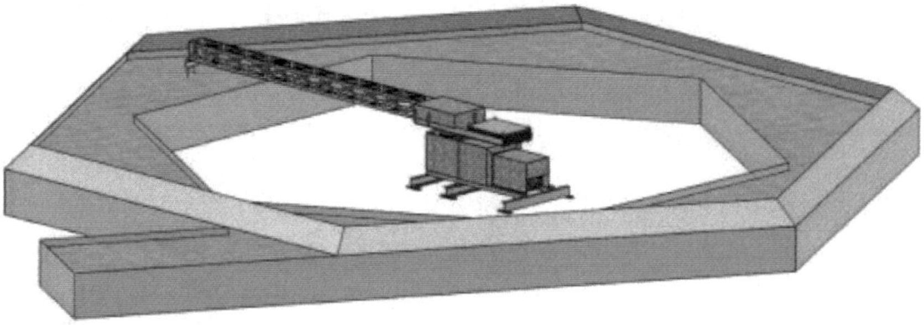

Figure 1. Illustration of the pool facility and the arm operated by the central electric motor for rotational motion control.

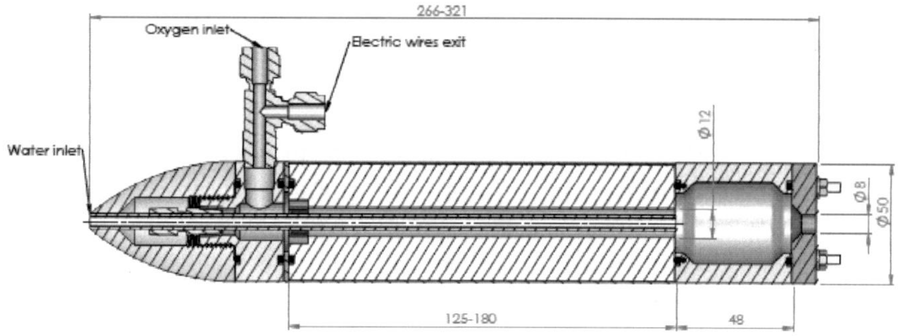

Figure 2. Section view sketch of the tested water breathing hybrid ramjet. Dimensions in mm

III. Results

Over 20 experiments have been conducted under various conditions, generating a performance map of the motor. Figure 3 shows the standard specific impulse (I_{sp}) measured versus the water-to-propellant mass ratio (W/P) for a specific oxygen-to-fuel mass ratio (O/F) range, revealing a significant increase of about 80% in specific impulse compared to an equivalent hybrid rocket without water entry. Figure 4 displays the motor's oxygen mass flow rate, chamber pressure, and velocity versus time for a single typical experiment. Focusing on the motor pressure chamber, the sequence begins in "static" section, where the motor's ram pressure (highlighted in green) is lower than the motor chamber pressure. Once the motor velocity reaches sufficiently high level, i.e., the ram pressure exceeds the motor pressure, water flows into the motor through the front inlet, causing pressure to increase until it reaches stagnation pressure at the desired velocity. This results in a significant increase in motor thrust.

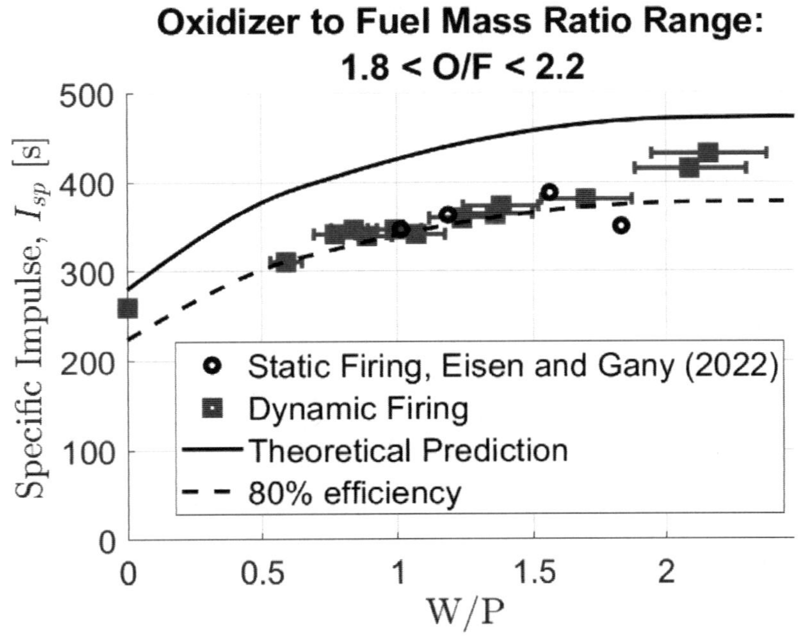

Figure 3. Dynamic firing test data of standard specific impulse (I_{sp}) vs water-to-propellant mass ratio (W/P) at oxidizer-to-fuel mass ratio (O/F) in the range of 1.8-2.2 compared to static firings [1] and theoretical prediction.

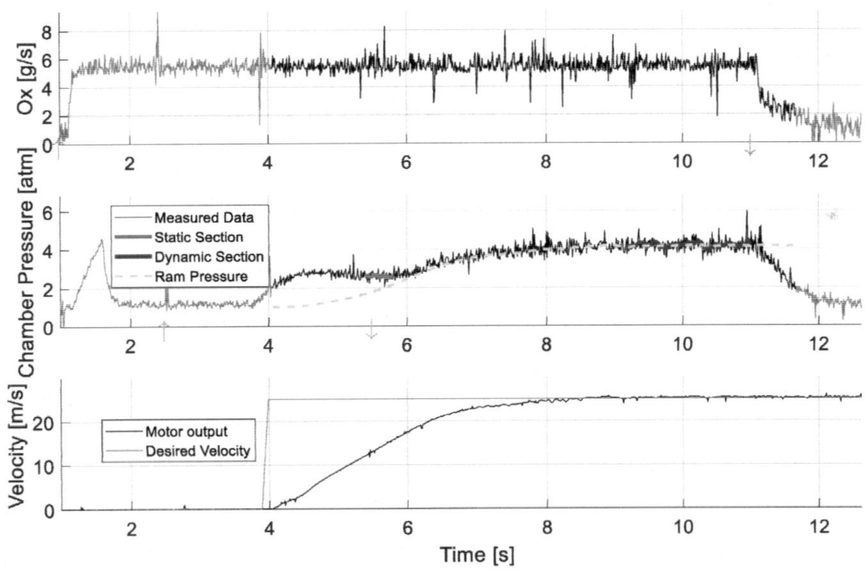

Figure 4. Oxygen mass flow rate (up), chamber pressure (middle) and motor velocity (down) vs time, for a single experiment at O/F ratio of 1.8 and W/P of 0.6.

References

[1] Eisen, N. E., and Gany, A., "Investigation of a marine water-breathing hybrid ram-rocket motor," *Journal of Propulsion and Power*, Vol. 38, No. 3, 2022, pp. 370–377. https://doi.org/10.2514/1.B38590

An exploration of wall-based ignition using nanosecond-pulsed high-frequency discharges

Weronika P. Senior - Tybora[1] and Joseph K. Lefkowitz[1]

Abstract— **Combustion in high-speed systems presents unique challenges, primarily due to the extremely short residence time of the gas in the combustion chamber. Combined with a highly turbulent environment, this often leads to rapid quenching of the ignition kernel, thus preventing successful ignition events. The nanosecond-pulsed high-frequency discharges (NPHFD) technique has emerged as a promising solution for addressing these challenges. NPHFD stands out for its ability to finely tune critical ignition parameters, such as the Pulse Repetition Frequency (PRF), Number of Pulses (N), and discharge voltage, which allows for control over the quantity and duration of energy deposition. This study aims to conduct a parametric exploration of the dynamics of wall-based ignition in flowing mixtures of ethylene (C_2H_4) and air using the NPHFD method. The primary objectives are to examine the ignition probability (P_{ig}) under high-speed and high-turbulence conditions similar to those in a scramjet cavity and to determine the Minimum Ignition Energy. The fuel-air mixture at 10 m/s is ignited by a burst of high-frequency discharges within a PRF ranging between 1-200 kHz. Two equivalence ratios of 0.45 and 0.5 were chosen, and the P_{ig} was studied as a function of PRF. Additionally, two different temperature setpoints, T_0=293K and T_1=373K, were selected, and their influence on the Pig was explored. The kernel was captured using a high-speed Schlieren imaging system and a high-speed IR camera equipped with a CO_2 filter. Results show that the higher the PRF, the easier and more robust the ignition, and a regime called a fully-coupled regime could be defined, where P_{ig}=1. The lower the PRF, the harder it is to ignite until no ignition is observed at 1 kHz. For the two temperature levels studied, it was found that higher temperatures achieved a higher Pig, accompanied by a broadened fully-coupled regime. Through this analysis, a deeper understanding of the NPHFD ignition process and its characteristics has been obtained. The impacts of PRF, equivalence ratio, temperature variation, and flow velocity are investigated in terms of P_{ig}, to better understand which set of conditions helps to achieve the highest P_{ig} overall.**

I. INTRODUCTION

Combustion is a complex process that can pose a challenge, particularly in the case of high-speed systems. The residence time of the gas in the chamber is of the order of milliseconds, meaning that a successful ignition has to be established in a short period of time. The highly turbulent environment can accelerate kernel quenching, potentially preventing ignition. One method demonstrating promising results in addressing ignition limits[1] and reducing ignition delay time is the nanosecond pulsed high-frequency discharge (NPHFD) technique[2]. Characterized by high voltages typically in the tens of kilovolts, discharge durations ranging from 1 ns to 100 ns, and a pulse repetition frequency (PRF) of up to 200 kHz, NPHFD allows for precise control over the level and duration of energy deposition through adjustments in discharge voltage, PRF, and the number of pulses (N). The NPHFD technique allows for accurate control of the level and the duration of energy deposition by adjusting the discharge voltage, PRF, and N (number of pulses). This flexibility enables efficient utilization of fluid motion to promote ignition kernel growth while minimizing energy and power requirements from the ignition source[3]. A study by Lefkowitz et al.[3] sidentified three inter-pulse coupling regimes: Fully-coupled, Partially-coupled, and Decoupled, depicted in Fig. 1. It was found that at short IPT (inter-pulse time) $< 2 \times 10^{-5}$ s P_{ig} (Ignition probability) was always equal one, and therefore the regime is called fully-coupled. For intermediate IPT - $\tau = 2 \times 10^{-4}$ s, the P_{ig} drops significantly; therefore, this regime is called partially-coupled. Lastly, a decoupled regime was defined corresponding to long IPT of the order of $\tau = 1 \times 10^{-3}$ s and P_{ig} was defined as a function of $P_{ig,signle}$ and N. Improving combustion for high-speed systems faces significant challenges which includes achieving a reliable ignition and sustaining a flame. The fundamental complexities of these conditions, characterized by high velocities, pose obstacles to conventional combustion processes. Traditional solution includes mechanical flameholders e.g. a wall cavity[4], [5], [6], [7] that allows to decrease the flow velocity and turbulence level which subsequently makes the air/fuel mixture easier to ignite. However, reaching a P_{ig} of 1 within a cavity remains challenging. Given the improved results with

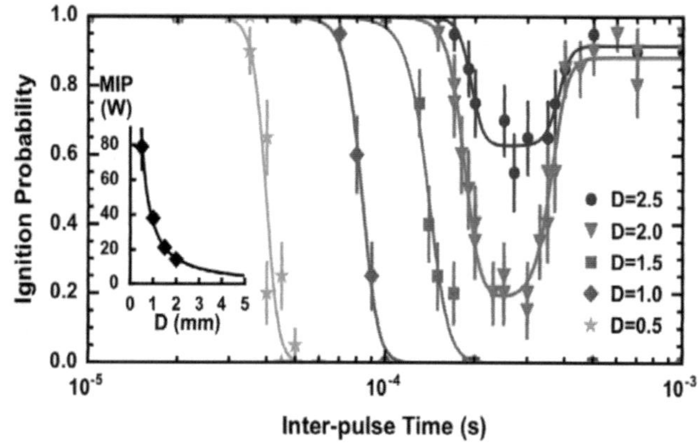

Fig. 1. Ignition probabiity as a function of IPT (s)

*This work was not supported by any organization

[1]Faculty of Aerospace Engineering, Technion - - Israel Institute of Technology, Haifa, 3200003, Israel weronikas@campus.technion.ac.il

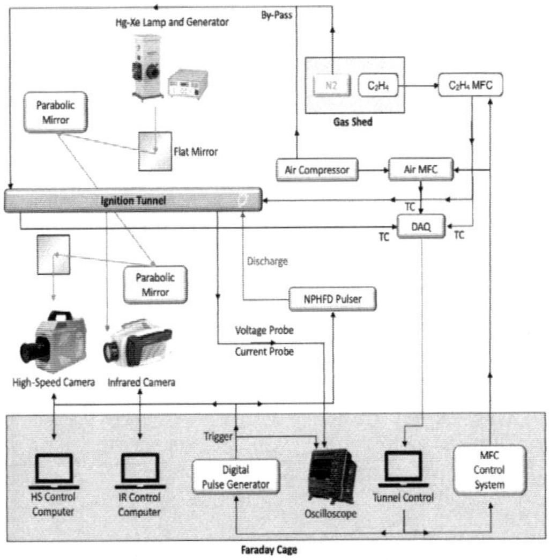

Fig. 2. Experimental setup

NPHFD compared to conventional discharges, it is believed that this technique could effectively address the complexities of fluid flow, enhancing ignition reliability. The objective of this study is to investigate the advantages of inter-pulse coupling within a controlled setting, assessing the probability of ignition in wall-based geometries as the PRF and equivalence ratio are varied.

II. EXPERIMENTAL SETUP

Experiments are conducted in a specially designed wind tunnel to create well-defined flow conditions for ignition-related investigations. The experimental system schematic is depicted in Fig.2. The setup consists of a stainless steel wind tunnel with a cross-section of 5 cm x 5 cm. Three mass flow controllers (MFCs) are installed. F-202AV-M20-RGD-22-V controls the fuel flow rate, and two Bronkhorst F-203AV-1M0-RGD-55-V controls the air flow rate. The flow rate of the air employed in the experiments oscillates around Q = 1366 L/min depending on the equivalence ratio, which corresponds to the fuel/air mixture velocity of U = 10 m/s. A high-flow rate air compressor powers the air supply. Additionally, a custom-built NPHFD pulser developed by FID GmbH is used. The pulse generator is triggered by a digital delay generator (Berkeley Nucleonics Corporation, Model 577). The NPHFD pulse generator's maximum PRF can reach up to 200 kHz, its maximum discharge voltage up to 20 kV, and its maximum discharge power up to 500 W. A high-voltage probe (Testec HVP 2739) and a current probe (Magnelab CT-E2.5-BNC) are employed to measure the discharge voltage and the current. Probes are situated within 1m of the discharge point due to the physical limitations. An oscilloscope (Teledyne Lecroy, Model HDO6104) records the discharge voltage and current waveforms. The following formula is used to compute energy from the voltage and current values:

$$\varepsilon = \int IV dt \tag{1}$$

The flow is introduced into the tunnel from one side and passed through a honeycomb structure with a hole diameter of 2.5 mm to dissipate large turbulence vortices and eliminate the possibility of a flashback. The experiments are conducted using a mixture of C_2H_4 and air for ϕ varying from 0.3 to 0.5. Valves, flow rates, PRF and N are controlled remotely with an in-home designed LabView program. A high-frame-rate Schlieren system, in a "folded-Z" configuration, facilitates the optical measurement of flame kernels utilizing two sapphire windows positioned on opposite sides of the tunnel. The system components include a 200 W continuous arc lamp (Oriel Instruments 66,477-200HXF-R1), two aluminum-coated parabolic mirrors with a 20.3 cm diameter and an effective focal length of 1.016 m, two silver-coated flat mirrors measuring 7.5 cm, a knife-edge, and a high-speed camera (Photron Fastcam SA-Z) operating at 50,000 Hz with a shutter speed of 0.36 ns. All ignition experiments include a single non-resistive automotive spark plug, positioned 86 mm downstream of the honeycombs. The selection of a spark plug featuring a surface discharge mechanism is deliberate, primarily aiming to minimize any disruptions within the flow field. For heated flow conditions two heaters (Tutco F077085, 36 kW) are used.

III. RESULTS & DISCUSSION

A. *Study of Ignition Limits*

NPHFD ignition limits were explored with frequency of 200 kHz and N = 100. Experimental trials were at first conducted under standard room conditions of P_0 = 100 kPa and T_0 = 297 K, and were followed by a campaign of heated flow conditions where temperature was increased to T_1 = 373.15. The flow speed used for the experiments was 10 m/s. Tests were conducted for multiple values of ϕ with 21 trials for each condition. Starting from an initial ratio of ϕ = 0.6, a systematic reduction of 0.05 in equivalence ratio was implemented until no ignition was observed at ϕ = 0.3, for both heated and non-heated cases, as shown in Fig.3. For $\phi >$ 0.4, there's a strong increase in P_{ig} values. P_{ig} was 1 for $\phi \geq$ 0.5 and 0.9 for ϕ = 0.45 . The results show that employing the NPHFD technique with high enough frequency allows us to achieve a successful ignition even for very lean cases. In the further approach a higher temperature was used and its influence on altering the P_{ig} was explored. The results demonstrated that an increase in temperature leads to higher P_{ig} values and allows for ignition across a wider range of equivalence ratios.

B. *Study of Ignition Probability*

Subsequently, the relationship between P_{ig} and IPT was explored. All experiments were conducted with a fixed number of pulses (N = 100), and two equivalence ratios of ϕ = 0.45 and ϕ = 0.5. Flow velocity was maintained at 10 m/s. The frequency of the NPHFD system was varied from 200 kHz to 1 kHz until the P_{ig} reached 0. Fig.4 illustrates P_{ig} as a function of PRF and corresponding IPT values. From the results , three regimes could be identified: fully-coupled, partially-coupled and decoupled. The fully-coupled

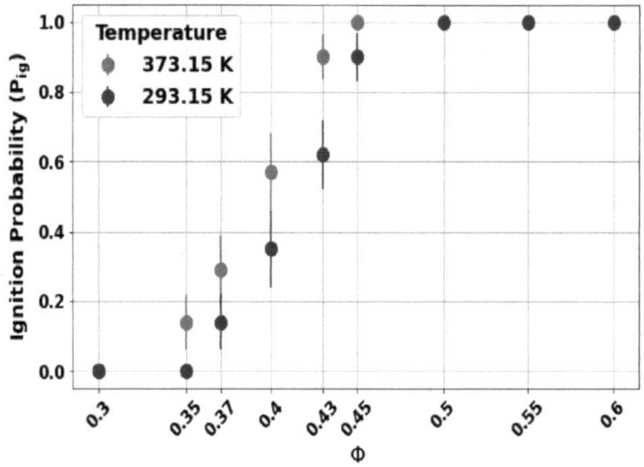

Fig. 3. Ignition probability as a function of equivalence ratio

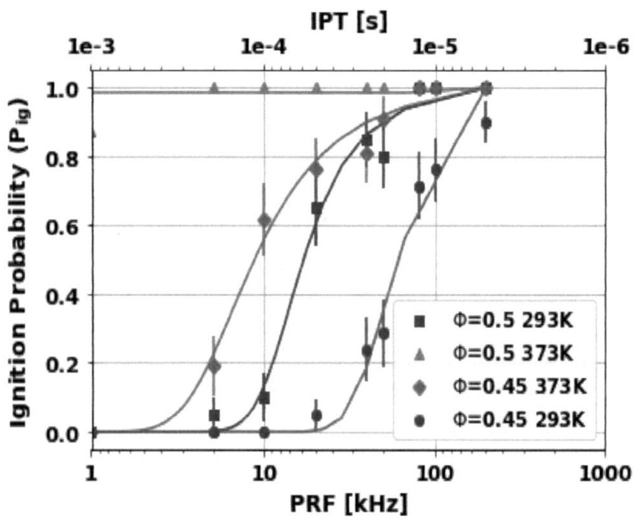

Fig. 4. Ignition probability as a function of PRF & IPT

regime could be identified for frequencies ranging from 80 kHz (IPT = 0.0125 ms) to 200 kHz (IPT = 0.005 ms). Subsequently, a decrease in P_{ig} was observed, and at 1 kHz, no ignition event was recorded. The error bars on the graph represent standard deviation values calculated based on P_{ig} measurements. One possible explanation for the difference in the partially-coupled regime between the center-based and wall-based ignition is the unique discharge geometry of the wall-based ignition. Hydrodynamic interactions between kernels in center-based ignition can lead to decreased values of P_{ig} in the partially-coupled regime. However, in wall-based ignition, the discharge does not act as a barrier to the flow, allowing the turbulent flow to follow the discharge without potentially destroying the previous ignition kernel. Another contributing factor could be the choice of fuel. C_2H_4 has broader ignition limits and is easier to ignite, making it more resistant to extinguishment even in the presence of potentially destructive interactions between subsequent ignition kernels.

IV. CONCLUSION

This study has effectively demonstrated the influence of NPHFD method on ignition probability in wall-based ignition systems under controlled conditions. By varying the pulse repetition frequency (PRF) and equivalence ratios, 3 inter-coupled regimes fully-coupled, partially-coupled, and decoupled were identified. The findings highlight the critical role of PRF in achieving and sustaining a successful ignition, particularly in systems where flow dynamics and ignition kernel interactions can significantly impact performance. The use of ethylene as a fuel showcased its beneficial properties in terms of resilience against extinguishment. These insights pave the way for optimizing ignition systems in high-speed applications, potentially leading to more efficient and reliable combustion.

ACKNOWLEDGMENT

This work is supported by The Israeli Ministry of Defense Directorate of Defense Research and Development.

REFERENCES

[1] D. A. Xu, D. A. Lacoste, and C. O. Laux. Ignition of Quiescent Lean Propane–Air Mixtures at High Pressure by Nanosecond Repetitively Pulsed Discharges. *Plasma Chemistry and Plasma Processing*, 36(1):309–327, January 2016.

[2] Sergey V. Pancheshnyi, Deanna A. Lacoste, Anne Bourdon, and Christophe O. Laux. Ignition of propane-air mixtures by a repetitively pulsed nanosecond discharge. *IEEE Transactions on Plasma Science*, 34(6):2478–2487, December 2006.

[3] Joseph K. Lefkowitz and Timothy Ombrello. An exploration of inter-pulse coupling in nanosecond pulsed high frequency discharge ignition. *Combustion and Flame*, 180:136–147, June 2017.

[4] Gautam Choubey and K. M. Pandey. Effect of variation of inlet boundary conditions on the combustion flow-field of a typical double cavity scramjet combustor. *International Journal of Hydrogen Energy*, 43(16):8139–8151, April 2018.

[5] Obula Reddy Kummitha, Lakka Suneetha, and K. M. Pandey. Numerical analysis of scramjet combustor with innovative strut and fuel injection techniques. *International Journal of Hydrogen Energy*, 42(15):10524–10535, April 2017.

[6] Peter Gerlinger, Peter Stoll, Markus Kindler, Fernando Schneider, and Manfred Aigner. Numerical investigation of mixing and combustion enhancement in supersonic combustors by strut induced streamwise vorticity. *Aerospace Science and Technology*, 12(2):159–168, March 2008.

[7] Gautam Choubey, Yuvarajan Devarajan, Wei Huang, Kulmani Mehar, Manvendra Tiwari, and K. M. Pandey. Recent advances in cavity-based scramjet engine- a brief review. *International Journal of Hydrogen Energy*, 44(26):13895–13909, May 2019.

Investigation of Radial Inward Porous Inert Media combustor for $NH_3/CH_4/H_2$ Combustion

G Mahesh Nayak*, Silky Elanjickal, Beni Cukurel, and Joesph Lefkowitz[1]

***Abstract*— An experimental study is conducted on a radial inward flow porous media burner (PMB) for ammonia (NH3)-air combustion. In this flow configuration, the cylinder surface area increases with its radius, and its velocity distribution is inversely proportional to the radius. As the flow travels through the porous media, it accelerates and is able to find a location at which it tends to stabilize as the kinematic balance between the burning velocity and the fresh mixture velocity are equal. The unique features of porous media enhance heat, mass, and momentum transfer, leading to an increase in the burning rate and improved stabilization of premixed flames to extend the flammable limit. Accordingly, radial PMBs offer a wider range of flame stability and power modulation rate compared to cylindrical axial burners, which are only dynamically balanced in a narrow range of flow rates. The present study utilizes a graded SiSiC Voronai-based lattice porous structure with a linear pore density of 10 to 20 PPI (pores per inch), progressing from the outer to inner radial sections. For various equivalence ratios and mass flow rates, temperature measurements conducted across various points of the combustion systems, including inlet mixture, pre-heated reactant mixture (before the porous media), and exhaust, are analyzed to understand the combustion characteristics of ammonia-air flames. In addition, exhaust emissions such as NO and unburned H_2 are discussed in detail.**

I. INTRODUCTION

Ammonia is a viable energy storage medium and has attracted significant interest recently. The significant challenges for using renewable ammonia in gas turbines are high NO_x and unburnt NH_3 emissions, poor combustion stability due to low flame speed ($\approx$ 7 cm/s), and resistance to ignition. Its adoption has been limited by its challenging combustion characteristics, resulting in poor efficiency and high NO_x and unburned NH_3 emissions. Numerous studies have been carried out to investigate the stable and efficient NH_3 combustion [1], [2]. Most of the studies reported hybrid combustion with hydrogen as a viable option due to the ability to reform NH_3 into H_2 on-site, as well as the extremely high flame speeds and flammability limits of H_2 as a promoter for NH_3 combustion. Li et al.[3] studied the improved burning velocity of NH_3 with the addition of H_2 for different equivalence ratios. The reported thermal NO_x of NH_3-H_2-air is reduced compared to pure H_2/air combustion. H_2, as an oxidant reduces the NO_x emission and improves the combustion performance of rich NH_3/air flames, tends to enhance flame sensitivity to flame stretching. Experimental studies have been reported on the swirl stabilized combustion of NH_3/air for various equivalence ratios, extending it further with the addition of 30% H_2 [4]. This NH_3/H_2 mixture is considered in many studies, as higher blends of H_2 ($>30\%$) lead to flashback and thermal NO_x. The unburned NH_3 and NO_x emissions are dependent on the NH_3/H_2 blends and equivalence ratios, and there is no clear trend in reducing NO_x [5]. The fuel-rich mixture of NH_3 is often reported to produce the lowest NO_x emission, accompanied by unburned reactants. It should be noted that the NO_x emission from NH_3 combustion is mainly due to fuel-bound nitrogen and incomplete combustion.

Recent studies investigated the PMBs concept for the NH_3 flames, which has not been extensively studied in the literature [2]. Nozari et al.[6] experimentally studied the NH_3/H_2/air combustion in silicon carbide (SiC) porous media burner to establish flammable limits for various mixture compositions to measure NO_x emission and combustion efficiency. However, the flame stabilizes at outside the ceramic foam indicating incomplete combustion at several operating conditions. An interface alumina–zirconia PMB is used to study the combustion of NH_3/H_2/air and NH_3/CH_4/air flame to evaluate the kinetic mechanism of NH_3 using experimental measurements [7]. Vignat et al.[8] demonstrated an interface zone of SiC-zirconia PMBs to stabilize NH_3/H_2/air flames for a wide range of equivalence ratios, mixture ratios, and mass fluxes. The NH_3 and H_2 ratio is found to be dependent on the equivalence ratio. It was reported that the NO emissions reach a minimum on the extinction limit conditions. They extended their study on the flame stabilization of NH_3/H_2/air and pollutant formation in a three-zone of PMBs [9]. The experimental measurements were analyzed with a reduced-order network model. In their study, rich and lean conditions of NH_3/H_2 ratio are found to operate at lower NO emission. The preheating of reactants increases the laminar speed of NH_3/H_2/air flame, given heat circulation and excess enthalpy combustion in PMB's provides a solution to NH_3.

Most of the previous studies investigated the axial cylindrical interface zone PMBs to stabilize the premixed flames. Very few studies reported in the radial flow burner [10], [11], [12]. The flame stabilization occurs at the kinematic balance between the burning velocity and fresh mixture velocity. Accordingly, radial PMBs offer a wide range of flame stability and rate of power modulation than cylindrical axial burners where velocity remains constant. The objective of the present study is to investigate NH_3/CH_4/ and NH_3/H_2 fuel mixture dynamics flame stabilization using an infrared imaging inside a solid matrix and exhaust emissions.

*[1] Authors with Faculty of Aerospace Engineering, Technion-Israel Institute of Technology, Haifa, Israel, 3200003.

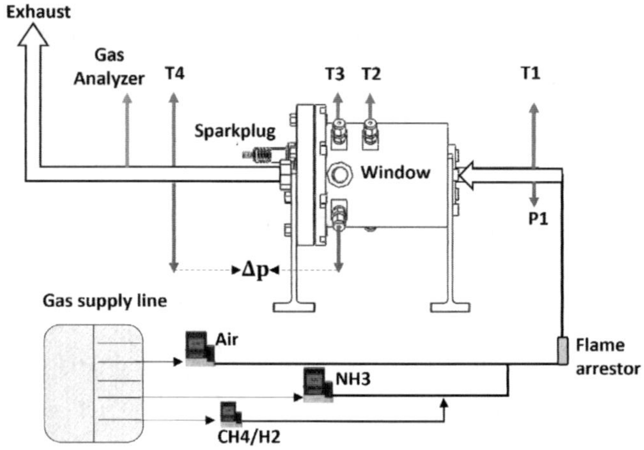

(a)

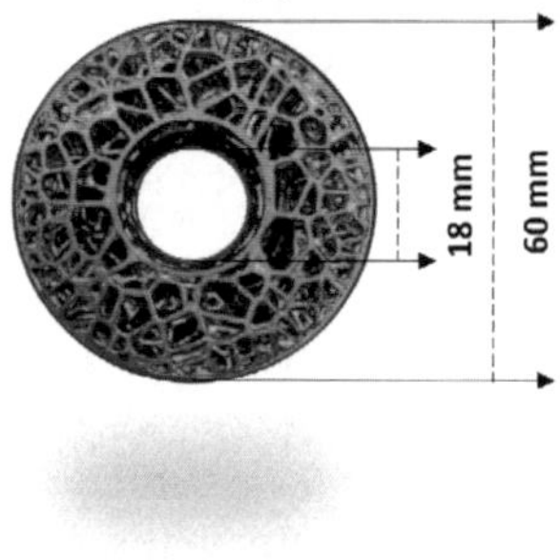

(b)

Fig. 1: (a) A schematic of experimental setup, (b) SiSiC voronoi lattice structure of porous media.

II. EXPERIMENTAL METHOD

A lab-scale combustor with mixtures of $NH_3/CH_4/Air$ and $NH_3/H_2/Air$ is examined in PMB. A schematic of the experimental setup is shown in Fig.1a. This study uses a cylidrical foam block made of SiSiC porous media with a height of 22 mm and an outer diameter of 60 mm. The inner diameters of 12 and 18 mm are employed for $NH_3/CH_4/Air$ and $NH_3/H_2/Air$, respectively, in Fig.1b. Both ceramic foams are graded linearly with 20-10 pores per inch (PPI) in a lattice voronois lattice structure with an approximate porosity of 0.85. The foam block is placed horizontally with an insulating material on both sides to maintain the inline direction of the combustor. The combustor is designed in such a way that the inlet mixture goes through outer circumference of cylindrical foam. Within a graded porous medium, the combustion is anticipated to stabilize by providing a geometrical advantage to alter the inlet mixture velocity. An array of mass flow controller regulates the reactant mixture of each gas. The experiments are conducted at room temperature and atmospheric pressure. An NI cDAQ control system consisting of four thermocouples is used to instrument the experimental setup. The temperatures of the inlet, combustor chamber, preheated mixture,and exhaust are recorded as T1, T2, T3, and T4, accordingly in Fig.1a. A gas analyzer (Testo -340) setup is incorporated into the system to characterize the NO emissions. Using gas chromatography, the unburnt H2 is reported only for the $NH_3/H_2/Air$ mixture. The reported emissions are repeated twice for each condition with an uncertainty of ±10% in the measuring range.

III. RESULTS AND DISCUSSION

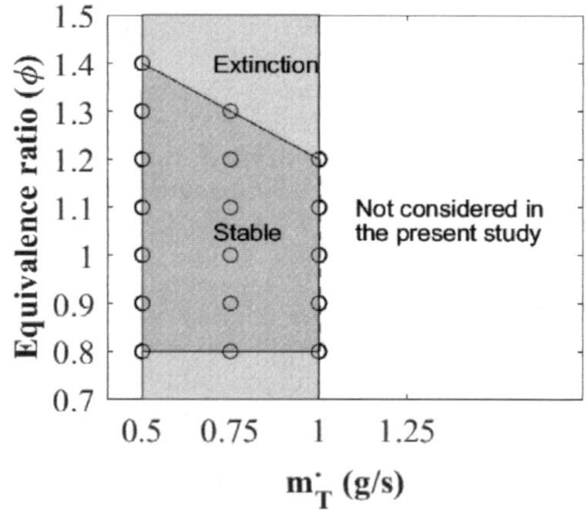

(a) NH_3/CH_4

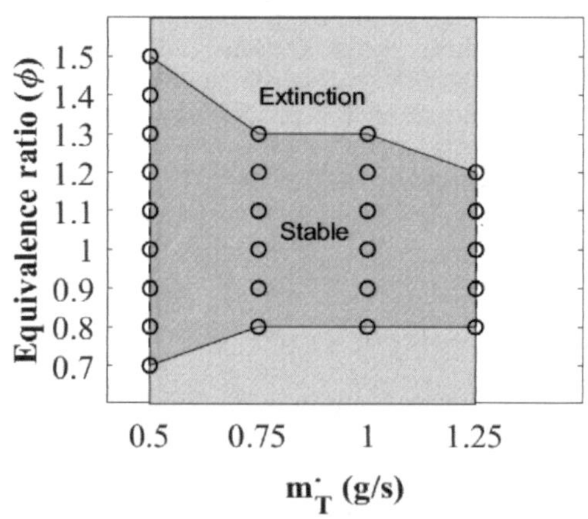

(b) NH_3/H_2

Fig. 2: Expperimental stability map for NH_3 mixtures.

Figure 2 displays the NH_3 mixture's combustion operational limits. The current experiments conducted a parametric study by varying the mass flow rate and equivalence ratio for a mixture of NH_3/CH_4 and NH_3/H_2 (a volumetric ratio of 90:10). The inward flow directed cylindrical foam is focused on the lean and rich conditions of two different chemical mixtures. The total mass flow rate is limited due to the high temperature of the exhaust. The combustion products from the inner circumference of foam are directed into the stainless steel tube, which is limited to extreme hot conditions. By maintaining the equivalence ratio and varying the mass flow rate, stable and extinction limits are found in this study.

Very few investigations have reported on the 90% NH_3 combustible mixture at the interface of two different porous media in an axial flow. The current porous media structure allows for a wide range of stable conditions for NH_3 with CH_4 and H_2 mixtures. The NH_3/H_2 mixture exhibits

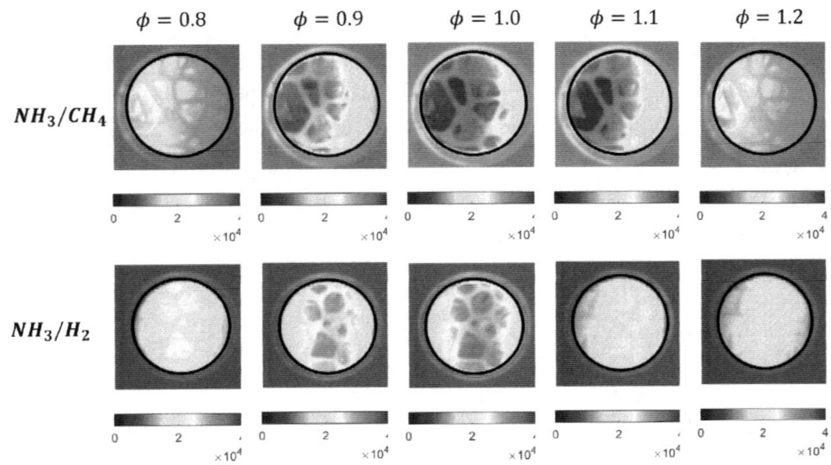

Fig. 3: Infrared H_2O emitted intensity inside a solid matrix of mass flow rate 1g/s for for a various equivalence ratios.

a longer, richer flammability than the NH_3/CH_4 mixture. These mixtures are influenced by the properties of their respective fuel components, CH_4 offers stability and predictable combustion behavior, while H_2 provides high reactivity and fast combustion kinetics. Lower mass flow rates result in more stable conditions. The lean extinction limit is seen at $\phi = 0.8$ for mass flow rates ranging from $\dot{m} = 0.5$ to 1 g/s. For NH_3/H_2, the lean and rich limits extend to $0.7 \leqslant \phi \leqslant 1.5$. The rich flammability limit is shown to fall as mass flow increases. The stable condition is defined as having a constant exhaust temperature within a margin of 5 K. Extinction points are monitored for two minutes before the flame is extinguished. It should be emphasized that the flashback operating points are not observed in this study due to the geometrical advantage of porous media. With the present configuration, a stable condition could be achieved with a pure NH_3/Air mixture at lower mass flow rates.

Image acquisition is carried out in the experiments through an optical window with a diameter of 16.5 mm. The window is positioned so that it provides radial access to the porous media. It is well known that the flame stabilizes at the interface of different porous media blocks in an axial flow. However, in a radial-directed reactant mixture, flame stabilizes inside the solid matrix at different feasible points. Therefore, it is crucial to determine the anchored flame location of the reactant mixtures. The current study uses an infrared camera (Telops FAST M3K) to detect broadband infrared emissions within a solid matrix. The image sequence is collected under a stable operating condition of one minute at a frame rate of 100 Hz. It should be noted that NH_3/CH_4 and NH_3/H_2 have different exposure durations, therefore, the emission intensity cannot be compared. Furthermore, porous media blocks vary in size with inner radius, although they represent a similar lattice structure. Figure 3 depicts a series of images from porous media combustion at different equivalence ratios of mass flow rate 1 g/s. The infrared image intensity of the combustion product H_20 can be seen for the reactant mixtures NH_3/CH_4 and $NH_3/H2$, respectively. The circular black outlines the size of the window covering the outer radius of the foam (right side) , in which the reactant mixture is radially supplied into porous media. Such a configuration provides a natural combustion flame front inside a solid matrix due to the alteration of inlet speed. It is significant for porous media combustion because the maximum combustion temperature conducts across the porous media, determining the exhaust emissions. The authors are not familiar with any reports showing combustion stabilization within a solid matrix. The qualitative images display the combustion flame front at different radial directions, demonstrating the geometrical benefit of porous media in this study. The lean and rich mixtures flame stabilizes downstream, whereas the stoichiometric mixtures stabilize more intensely upstream. It is evident that the flame front propagates relative to the fuel-air mixture within the solid matrix. Furthermore, the combustion front moves downstream in accordance with mass flow rates at a constant equivalence ratio. Future studies will be conducted to compare both mixtures under similar conditions.

Figure 4 and 5 show the measured temperature and emission for the NH_3/CH_4 and NH_3/H_2 fuel mixtures, as a function of equivalence ratio and mass flow rates, respectively. The exhaust temperature is measured at the location of T4 as shown in Fig.1a. The reported temperature is not corrected for the radiation losses of the exhaust tube. The infrared images depict the low and high temperature regions across the porous media. It clearly elucidates the radial temperature gradient in the solid matrix zone, in which the flame front in NH_3/H_2 is earlier than the NH_3/CH_4 fuel mixture due to the high reactivity of the H_2. The exhaust temperatures of NH_3/CH_4 and NH_3/H_2 are more likely to be the same despite the fact that the thermal input is different. It can be addressed by increasing the radial size of the solid matrix in the NH_3/H_2 fuel mixture. It should be noted that the measured exhaust temperature is approximately the same throughout the equivalence ratio for a mass flow rate. Based on Fig.3, the temperature observed to be different with the equivalence ratio, where maximum temperature can be seen at stoichiometric. For an equivalence ratio, the

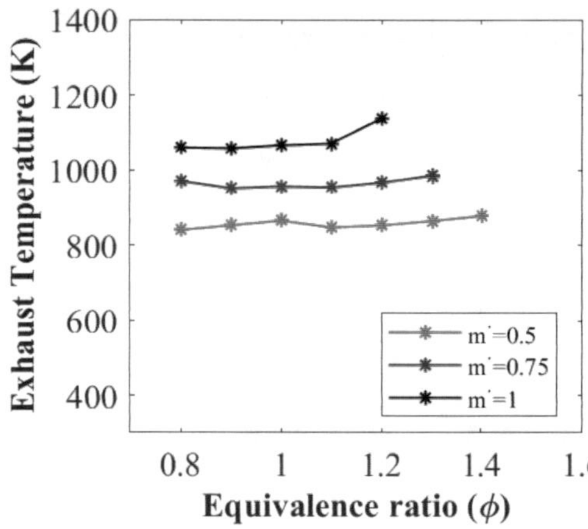

(a)

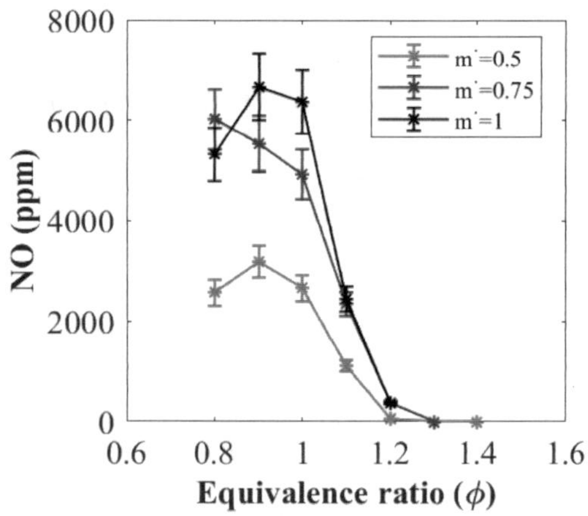

(b)

Fig. 4: NH_3/CH_4 fuel mixture, (a) Exhaust Temperature, (b) NO emission

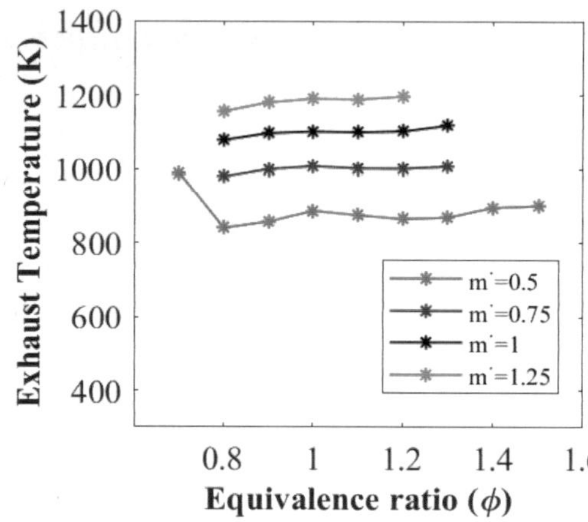

(a)

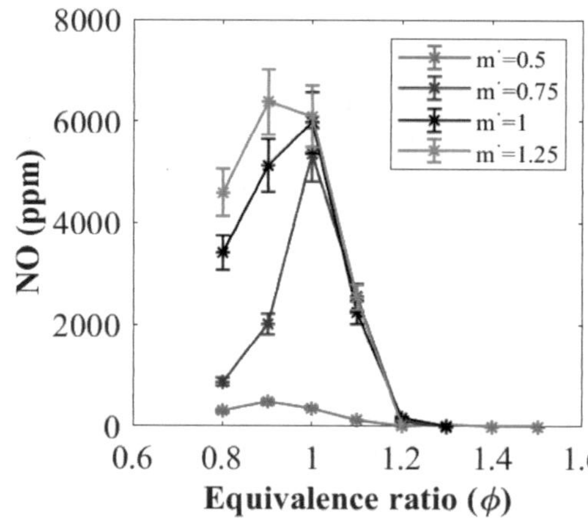

(b)

Fig. 5: NH_3/H_2 fuel mixture, (a) Exhaust Temperature, (b) NO emission

exhaust temperature rises in proportion to mass flow rate. In Fig.5a, the temperature at ϕ=0.7 for $\dot{m}$ =0.5 g/s is higher due to downstream flame stabilization, where there is less availability of heat conduction into the solid matrix.

Figure 4b and 5b depict the measured NO emission for the NH_3/CH_4 and NH_3/H_2 fuel mixtures as a function of equivalence ratio. The reported NO emissions are averaged over 1 minute. The exhaust emissions in both fuel mixtures ranged between 2000 and 6500 ppm. It has been previously reported that the consumption of NH_3 occurs mostly by H abstraction, with hydroxyl radicals (OH) being the primary reactants under all conditions of equivalence ratio. Other secondary consumption processes involve interactions with hydrogen (H) and oxygen (O), which result in the formation of the common product NH_2. The oxidation of NH_i species (where i denotes various nitrogen-hydrogen combinations) has two possible outcomes, depending on the proportion of O/H radicals. If the O/H radical ratio is large, NH_i species are more likely to be oxidized, resulting in the production of nitric oxide (NO). This process primarily involves the generation of an intermediate molecule known as HNO. On the other hand, as the concentration of O/H radicals decreases, NO reduction via NH_i+ and NO reactions becomes more dominant. In this investigation, significant NO emissions are detected for both fuel blends around the stoichiometric. The NO emissions are anticipated to develop at a high temperature zone in the solid matrix, as seen in Fig.3. The production rate of NO is higher when ϕ=0.9 for NH_3/CH_4 and ϕ=1 for NH_3/H_2 fuel mixtures. A 50% increase in mass flow rate from 0.5 g/s to 0.75 g/s for NH_3/CH_4 causes a one-order-of-magnitude increase in NO emission. However, with NH_3/H_2 fuel mixture, NO emissions are substantially lower at $\dot{m} = 0.5$ g/s, where NO is detected around 200 ppm. To investigate further, the availability of O_2 in the exhaust gases is shown in Fig.6a. The measured $O2$ concentration elucidates the equivalence ratio of the experimental conditions. At $\dot{m}$ =0.5 g/s, a substantial amount of O_2 in the exhaust explains the partial combustion caused by the transport rate of reactants. In addition, hypothetically, at lower temperatures of NH_3/H_2 mixture, combustion kinetics might differ from the other experimental conditions, hence the lower formation of NO.

An increase in mass flow results in elevated NO emissions, as reported in previous studies. The higher the flow rate, the more fuel is added to the combustion process, and thus the

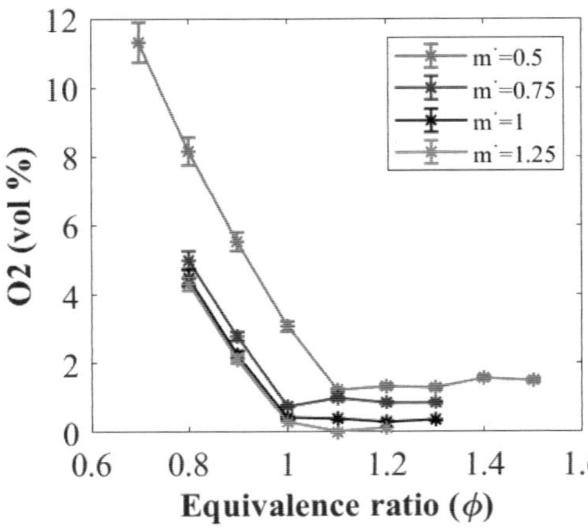

(a)

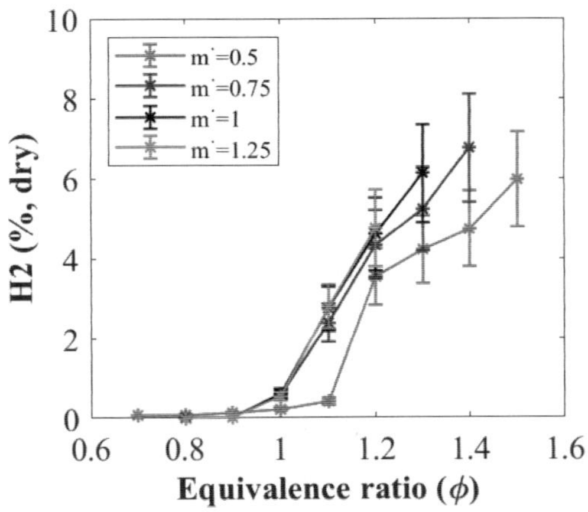

(b)

Fig. 6: Exhaust gas measurements of $NH_3/H2$, (a) Measured O_2, (b) Unburnt H_2

temperature rises, as do NO emissions. In both fuel mixtures, NO emissions are lower in rich conditions compared to the stoichiometric and continue to fall with the equivalence ratio. For rich mixture $\phi \geq 1.2$, NO emissions are on the order of 50 ppm, and no longer detectable by the gas analyzer. Simultaneously, the exhaust emissions contain an increasing concentration of unburned fuel. Richer combustion produces unburned fuel concentrations in the form of NH_3 and H_2, with a considerable fraction resulting in unburnt H_2. In this study, H_2 volume fraction gradually increased to 6% in Fig.6b. The H_2 concentration is found to increase with equivalence ratio and mass flow rate. It should be noted that the unburnt H_2 at $\dot{m}$ =0.5 g/s is in the similar range of other experimental conditions. It encourages to investigate the reaction pathways with detailed chemistry in this operating condition.

IV. CONCLUSIONS

The experimental study explored the combustion characteristics of NH_3 mixtures, particularly focusing on CH_4 and H_2 blends. Findings underscore the significance of cylindrical foam as a combustion medium, providing insights into flame stabilization mechanisms. The experiments successfully combusted 90% ammonia mixture, demonstrating its compatibility with both methane and hydrogen as co-reactants. Through infrared imaging, the current study identified distinct flame stabilization patterns inside the solid matrix, highlighting the geometric advantages of radial inward flow porous media burners. Temperature and emission measurements revealed the effect of mass flow rates and equivalence ratios on combustion, showing trends in NO emissions and unburnt fuel concentrations. Notably, at lower equivalence ratios, NO emissions tended to increase, peaking near stoichiometric conditions before falling in richer mixtures. Specifically, at a mass flow rate of 0.5 g/s, NO emissions exhibited unique characteristics, with NH_3/H_2 mixtures showing a notable decrease in NO levels compared to NH_3/CH_4 blends. This observation suggests the importance of detailed chemistry considerations and flow dynamics in understanding and mitigating NOx emissions in NH_3-based combustion systems.

REFERENCES

[1] A. Valera-Medina, F. Amer-Hatem, A. Azad, I. Dedoussi, M. De Joannon, R. Fernandes, P. Glarborg, H. Hashemi, X. He, S. Mashruk, *et al.*, "Review on ammonia as a potential fuel: from synthesis to economics," *Energy & Fuels*, vol. 35, no. 9, pp. 6964–7029, 2021.

[2] A. M. Elbaz, S. Wang, T. F. Guiberti, and W. L. Roberts, "Review on the recent advances on ammonia combustion from the fundamentals to the applications," *Fuel Communications*, p. 100053, 2022.

[3] J. Li, H. Huang, N. Kobayashi, Z. He, and Y. Nagai, "Study on using hydrogen and ammonia as fuels: Combustion characteristics and nox formation," *International journal of energy research*, vol. 38, no. 9, pp. 1214–1223, 2014.

[4] M. C. Franco, R. C. Rocha, M. Costa, and M. Yehia, "Characteristics of nh3/h2/air flames in a combustor fired by a swirl and bluff-body stabilized burner," *Proceedings of the Combustion Institute*, vol. 38, no. 4, pp. 5129–5138, 2021.

[5] A. Valera-Medina, D. Pugh, P. Marsh, G. Bulat, and P. Bowen, "Preliminary study on lean premixed combustion of ammonia-hydrogen for swirling gas turbine combustors," *International Journal of Hydrogen Energy*, vol. 42, no. 38, pp. 24 495–24 503, 2017.

[6] H. Nozari, G. Karaca, O. Tuncer, and A. Karabeyoglu, "Porous medium based burner for efficient and clean combustion of ammonia–hydrogen–air systems," *International journal of hydrogen energy*, vol. 42, no. 21, pp. 14 775–14 785, 2017.

[7] R. C. Rocha, C. F. Ramos, M. Costa, and X.-S. Bai, "Combustion of nh3/ch4/air and nh3/h2/air mixtures in a porous burner: experiments and kinetic modeling," *Energy & Fuels*, vol. 33, no. 12, pp. 12 767–12 780, 2019.

[8] G. Vignat, B. Akoush, E. R. Toro, E. Boigné, and M. Ihme, "Combustion of lean ammonia-hydrogen fuel blends in a porous media burner," *Proceedings of the Combustion Institute*, 2022.

[9] G. Vignat, T. Zirwes, E. R. Toro, K. Younes, E. Boigné, P. Muhunthan, L. Simitz, D. Trimis, and M. Ihme, "Experimental and numerical investigation of flame stabilization and pollutant formation in matrix stabilized ammonia-hydrogen combustion," *Combustion and Flame*, vol. 250, p. 112642, 2023.

[10] A. Mohamad, "Combustion in porous media: fundamentals and applications," in *Transport phenomena in porous media III*. Elsevier, 2005, pp. 287–304.

[11] C. Bedoya, I. Dinkov, P. Habisreuther, N. Zarzalis, H. Bockhorn, and P. Parthasarathy, "Experimental study, 1d volume-averaged calculations and 3d direct pore level simulations of the flame stabilization in porous inert media at elevated pressure," *Combustion and Flame*, vol. 162, no. 10, pp. 3740–3754, 2015.

[12] I. Yakovlev, A. Maznoy, and S. Zambalov, "Pore-scale study of complex flame stabilization phenomena in thin-layered radial porous burner," *Combustion and Flame*, vol. 231, p. 111468, 2021.

Dual Mode Scramjet Operation and Engine Unstart

Eran Arad

Mechanical Engineering, Technion - Israel Institute of Technology

Ido C Ruhman

Aeronautical Systems, Rafael Advanced Defense Systems LTD

Abstract

The current study focuses on a significant challenge in the application of Scramjet, specifically dual-mode combustion with a constant geometry. A strong interaction between the combustion chamber and inlet, particularly at the onset of thermal-choking, could lead to inlet unstart and engine failure[1].

The analysis methodology involves numerical simulation, performed at both low and high throttle settings within the flight envelope, and throughout the mode transition. This complete flow-path simulation allows us to assess the interaction and the strong coupling between the domains.

Simulations are conducted using compressible Large Eddy Simulation (LES), coupled with a Compressible Flamelet-Progress-Variable (CFPV) combustion model and a detailed chemical-kinetic mechanism[2]. This approach has been validated in previous publications[3,4] against measurements in the simulation of supersonic combustion in a combustion chamber, in a connected-pipe mode. Reproduction of a milestone hypersonic inlet experiment[5] was used for validation of the inlet and isolator flow.

We use a generic complete flow-path configuration, which includes an inlet, isolator, combustion chamber, flame holder, and nozzle, for the analysis. Both inertial flow through the entire flow path and ignition of supersonic combustion were investigated. Data was accumulated in the two combustion modes under consideration. Contour plots of the simulation results, obtained at flight level of 28Km, Mach number of 6 and $\alpha = 3^o$ are presented in figure 1.

(a) (b)

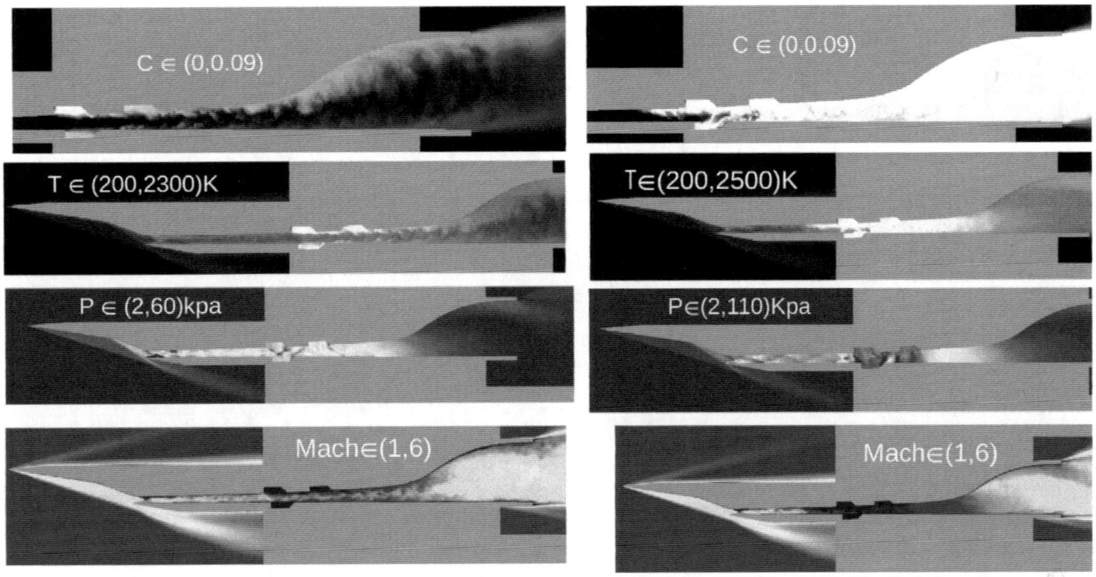

Figure 1: Contours plots of flow variables on a span-wise center crosscut. (a) ϕ=0.3 SCRAM mode; (b) ϕ=0.75: RAM mode. C is the Progress Variable

Tentative Conclusion and Outlook

The aim of this research is to provide guidelines for the proper coupling of inlet design with dual-mode combustion, to identify the conditions that may lead to inlet unstart, and to explore methodologies for smooth and proper operation through the mode transition. In the presentation we delve into diverse fuel mixture policies and their impact on ignition, combustion and propulsion dynamics.

References

1. Laurence S J, Lieber D, Schramm J M, Hannemann, Larsson J, "Incipient thermal choking and stable shock-train formation in the heat-release region of a scramjet combustor. Part I: Shock-tunnel experiments", Combustion and Flame 162 (2015) 921–931

2. Saghafian A, Shunn L , Philips D A and Ham F, "Large eddy simulations of the HIFiRE scramjet using a compressible flamelet/progress variable approach", Proceedings of the Combustion Institute 35 (2015) 2163–2172

3. Arad E, Yokev N, Brod H E and Michaels D,"From Supersonic Combustion to Thermal Choking: Numerical and Experimental Analysis of A Scramjet Engine", AIAA-2020-2966

4. Arad E and Michael D, "Analysis of Air Vitiation and Contamination Effects on Scramjet Ground Tests", AIAA-paper 2021-4177

5. Häberle J and Gülhan A, ”Investigation of Two-Dimensional Scramjet Inlet Flowfield at Mach 7", JOURNAL OF PROPULSION AND POWER, Vol. 24, No. 3, May–June 2008

Pulse Detonation Ignition of n-Dodecane Spray and Air Mixtures

Hertzel Kadosh[1] and Dan Michaels[2]

Faculty of Aerospace Engineering, Technion - Israel Institute of Technology, Haifa, Israel

Pulse detonation is an alternative method for energy addition and ignition of the fuel-air mixture in challenging conditions that exist in ramjet and scramjet combustors. This study investigates the ignition and flame spreading of a dodecane spray and air mixture with a pulse detonator operating with liquid acetone and oxygen. Experiments were conducted in a 3-meter-long rectangular channel with a fuel injection in its head and heated air at a pressure of 1 atm and an initial temperature of 298, 350, 400, and 450 K. A pulse detonation was applied 45 cm downstream of the channel inlet with a fill fraction equal to 1 or 2. The size distribution of dodecane fuel droplets was characterized in open space and through optical windows in the channel. A simultaneous Schlieren and chemiluminescence hi-speed imaging was used to image the flow and combustion. At an initial temperature of 350 K and lower, the pulse detonation did not lead to ignition and flame propagation in the channel, probably because of insufficient fuel vapor, resulted by wall wetting and low evaporation rate. Moreover, the mixture was auto-ignited at an initial temperature of 480 K. At air temperatures of 400 K and 450 K, different flame speeds were measured for different initial fill fractions, which had a dominant effect. A faster flame was observed for a higher fill fraction and initial temperature. At an initial temperature of 450 K and a fill fraction ff = 2, a mean flame speed of 315 m/s was measured at 145 cm downstream from the pulse detonator exit. Chemiluminescence images showed that a higher volume adjacent to the detonation tube exit is occupied by the flame and that it lasts for a longer time in that case. It is assumed that at a higher fill fraction, the detonation wave follows the path of the reactive mixture, proceeds further into the channel while decoupling from the leading shock wave, and delivers its high-temperature and high-pressure products to the mixture. That results in the engagement of a greater volume of mixture, increasing its ignition capability.

---

[1] Ph.D. student.
[2] Professor, Head, Fine rocket propulsion center.

Modular Modelling Approach for Efficient Gradient Evaluation of a Low-Order Integrated Scramjet Propulsion System

Amir Mittelman[1], Kieran Mackle[1], Ingo Jahn[2], Rowan Gollan[1]

Hypersonic scramjet-propelled vehicles are drawing attention from the worldwide space community as they present the potential for increasing operational flexibility while reducing launch costs. However, complex physical phenomena encountered in the hypersonic flight regime result in a highly coupled and constrained vehicle design space. As the design grows in complexity, more disciplines become involved in the optimization process, posing several challenges. First, the design space rapidly expands to include an ever-growing number of design variables. Second, different subsystems require different simulation approaches and tools, so using a single `tip-to-tail' simulation framework can be highly limiting. In the modular modelling approach, sub-system disciplinary models are joined together, allowing for efficient calculation of cross-coupled, large-scale system objectives (performance) and design sensitives (relationship between performance and design variables). In this work, we adopted the modular design framework approach and demonstrated the feasibility of efficiently yet accurately calculating the vehicle sensitivity matrix by combining quick-to-compute component-level partial derivatives. The vehicle design framework coupled together a newly developed low-order 2-D inlet and cowl, 1-D combustor analysis tool `HyperPro`, with the aerodynamic solver `PySAGAS`. An isentropic sensitivity calculation was added to the aerodynamic solver `PySAGAS` for nozzle sensitivities analysis. The system level performance and sensitivity calculations were verified by comparison to a 2-D inviscid CFD solver - this showed good accuracy of the modular and isentropic approach. Additionally, the use of sub-system partial derivatise demonstrated the capability of realizing system-level insights at the component-level design. The demonstrated effectiveness and flexibility confirm the potential of this approach for Multidisciplinary Design Analysis and Optimization (MDAO) when applied to complex aerodynamic and fluid dynamic systems such as scramjet-propelled hypersonic platforms.

[1] School of Mechanical and Mining Engineering, Centre for Hypersonics, University of Queensland, St. Lucia QLD 4072, Australia

[2] Institute for Advanced Engineering and Space Sciences, Springfield Campus, The University of Southern Queensland, Toowoomba, QLD, 4350, Australia

Helical Hybrid Engine Operating in a Blowdown Regime, Technion Rocketry Club

Ethan Loskove[1], Zvi Weston[2], Michal Amar, Noa Segal, Ohr Bershadsky, Yonatan Bason, Sagi Dinisman[3]

Technion Israel Institute of Technology, Haifa, 3200003, Israel

Nachum Eisen[4]

Israel Aerospace Industries, Lod, 7110604, Israel

The Technion Rocketry Club is a student run club whose aim is to launch an in-house student built and designed rocket. The propulsion team is responsible for the engine and its associated components.

We are utilizing a hybrid rocket engine for our rocket. Hybrid engines are composed of a fluid oxidizer and solid propellant and are advantageous to use. They are less mechanically complicated as compared to liquid fueled rockets and are easier/safer to produce and operate in comparison to solid fuel rockets.

In our hybrid engine we are using O2 gas as our oxidizer. While it is less common to utilize gaseous O2 in hybrid rockets, gaseous O2 simplifies our design and there is no need to incorporate more complicated oxidizer injection schemes. The solid fuel we chose is Polyester as it is an advantageous fuel. Polyester is inexpensive, safe to store and experiment with, easily castable into designed fuel grain geometry, and has high performance characteristics [1].

The shape of the fuel grain we chose to be helical to increase the burn area. The design parameters of the fuel grain allow us to achieve the desired mass flow rates in conjunction with the stoichiometric peak for our propellants. In addition, the helix allows the desired flow rates to be achieved in a smaller fuel grain, and therefore enables our engine to be more volumetrically efficient and less massive. Using previous data [2] and iterating though different grains, we were able to design our own grain geometry that produces the desired ballistics profile.

Our oxygen injection system operates as a blowdown rather than with regulated pressure. Working with a regulator would have added weight that decreases the overall performance of the rocket. The regression rate of the propellants changes rapidly due to the dramatic change of the oxygen mass flow rate in our blowdown scheme and the changing fuel grain geometry. We modeled the flow rates of the oxygen and fuel in the blowdown regime and arrived at a theoretical model for the thrust varying with time. We performed experiments to validate the theoretical model. In static fires tests (figure 1) we emulated a blowdown and arrived at an experimentally supported thrust curve for our hybrid engine in agreement with our theoretical model.

---

[1] Former Head of Propulsion Team, Faculties of Aerospace Engineering and Physics

[2] Current Head of Propulsion Team, Faculty of Mechanical Engineering

[3] PhD Student, Faculty of Aerospace Engineering

[4] PhD, Aerospace Engineer Missile and Space Group

Figure 1: Static fire experiment

References

[1] Eisen, N.E. and Gany, A. (2022) 'Investigation of a marine water-breathing hybrid ram-rocket motor', Journal of Propulsion and Power, 38(3), pp. 370–377. doi:10.2514/1.b38590.

[2] Dinisman, S., Eisen, N.E. and Gany, A. (2024) 'Enhancing hybrid motor thrust by a helical-port fuel grain', International Journal of Energetic Materials and Chemical Propulsion, 23(1), pp. 69–84. doi:10.1615/intjenergeticmaterialschemprop.2023049385.

Enhancing Elastic Wave Attenuation in Single-Phase Phononic Crystals via the Coupling of Bragg and Local Resonance Mechanisms

Ilaie Nadejde* and Pavel I. Galich†

Faculty of Aerospace Engineering, Technion–Israel Institute of Technology, Haifa, Israel

Abstract

Phononic crystals (PnCs) are periodic materials based on specific geometric designs that create a periodic impedance contrast. In single-phase PnCs, the impedance contrast occurs between the solid material and air. PnCs can attenuate elastic waves across various frequencies through their unique band structure, leading to the development of numerous acoustic devices, including waveguides, filters, and vibration isolation technologies. With advances in numerical tools and fabrication technologies, the challenge in research today lies in enhancing the wave attenuation performance of PnCs, broadening the forbidden frequency ranges, and minimizing their size. The phononic band structure of a PnC is typically determined by its geometry, mechanical properties, and symmetry. High symmetries increase the probability of achieving a complete omnidirectional phononic bandgap (BG) – a range of frequencies where waves are exponentially attenuated for all wave vectors due to Bragg scattering, local resonance (hybridization), or weak elastic coupling mechanisms. In our recent study [1], we exploited such higher symmetries to push the limits of complete omnidirectional BGs in 2D PnCs, unveiling an exceptionally wide BG of 124% for the nonsymmorphic p4gm symmetry group design. This design surpasses symmorphic p4mm-group PnC classic designs in both BG size and attenuation performance. Any 2D periodic structure can be categorized into one of 17 plane symmetry groups. Among these groups, four are nonsymmorphic (p4gm, p2mg, p2gg, and p1g1), and are defined by a glide operation as an irreducible symmetry element of the plane group. In this study, we explore how complex symmetries and geometric features enhance BGs in single-phase 2D PnCs by comparing their BG formation mechanisms. Specifically, we examine the causes that enable the p4gm design to outperform the classic p4mm design in terms of both BG size and attenuation. This is possible by the coupling of the Bragg scattering and local resonance mechanisms in the p4gm design. We numerically compute the BG for both p4mm and p4gm PnC designs and investigate how different designs and their corresponding mechanical properties influence BG formation mechanisms. The type and attenuation performance of the BG are determined by analyzing the vibration patterns and the imaginary parts of the Bloch wave vectors (evanescent modes), which provide quantifiable indicators of the waves' exponential decay. We conduct transmission loss simulations and experiments, for both P- and rarely explored S-waves, to confirm the improved wave attenuation and wide BG achieved by coupling the two BG mechanisms. Our findings shed light on the mechanisms contributing to wide BGs with enhanced wave attenuation, paving the path for improved and better designed acoustic devices.

*MSc Student, ilaienadejde@campus.technion.ac.il.

†Assistant Professor, galich.pi@technion.ac.il.

References

[1] Nadejde, I., Thomas, E. L., and Galich, P. I., "Pushing the limits of complete omnidirectional bandgaps in 2D nonsymmorphic single-phase phononic crystals," *Applied Physics Letters*, Vol. 123, No. 13, doi:10.1063/5.0157222.

Designing a multistable unit cell using bistable arches

Ankush Yadav and Pavel I. Galich

Faculty of Aerospace Engineering, Technion–Israel Institute of Technology, Haifa, Israel

Abstract

In recent years, multistable structures have gained a lot of attention due to their applications in fields such as energy dissipation, logic gates, surface morphing, and soft robotics. The concept of snap-through instability is commonly employed to create a bistable unit cell. These bistable unit cells, when arranged in 1D, 2D, or 3D arrays, can show multiple stable configurations (at least more than two stable states), thus, giving rise to multistable structures.

In this work, we present a design approach for creating multistable unit cells comprised of curved bistable elements (i.e., arches). Our multistable cell is modeled using commercial Finite Element software COMSOL 6.2. We assume the cell elements to be made from a neo-Hookean material. By introducing contrasts in the elastic moduli of the curved beam elements and fixing the unit cell at its four vertices, we demonstrate the emergence of up to five stable configurations within the unit cell. In one configuration, snap-through does not occur, while in others, distinct force-displacement responses are observed when loading is applied to the right side of the cell. This variation in response can help us tune the behavior of the unit cell without changing its structure. Additionally, we investigated how different configurations affect the stiffness of connecting arches and how geometric properties (such as an apex height of the curved arch) can alter the number of stable configurations and force responses of these configurations. When arranged in a 1D array, these multistable cells can support transition wave propagation, making them applicable in mechanical logic devices, such as soft robotics.

Arc discharge heating for high-enthalpy wind tunnel

One of the biggest challenges for atmosphere re-entry vehicles is surviving the extreme heat fluxes produced by friction with the surrounding air. Such vehicles require special thermal protection systems to ensure their survival. As real hypersonic vehicle flight tests are very expensive, it is necessary first to carry out tests in high-enthalpy facilities. One of the best ways to generate such high-enthalpy flow for a relatively long time (on the order of few minutes) is using an arc-heated wind tunnel. A plasma arc refers to a gas in the plasma state generated by high-voltage discharge through it. This hot plasma is then accelerated into a test chamber towards a target. There exist different types of discharges, where the two main ones are glow-discharge, and arc-discharge. Here we will examine the differences between the two discharge types and learn why an arc-discharge is used in high-enthalpy facilities. We will discuss the importance of correctly characterizing the plasma arc. In addition, we will describe the difficulties of diagnostics of such plasma and suggest solutions. These are based on specialized high-voltage and high-current measurements, emission spectroscopy, and numerical simulations.

Computed Tomography Integration into the PHFGMC Meso-Micro Mechanical Characterization of C/C-SiC Composites

Royi Padan, Chen Dahan-Sharhabani , Omri Regev and Rami Haj-Ali*
Tel Aviv University
Rafael Advanced Defence Systems Ltd.

Ceramic matrix composites (CMCs) are an ideal candidate for structural materials in space or hypersonic vehicles. Therefore, reliable simulation tools of their thermomechanical behavior are critical. CMCs, especially C/C-SiC, are highly affected by the microstructure of their local multi-phase constituents. Recent studies employed different micromechanical models to evaluate CMCs' material properties. However, they focused mainly on evaluating the elastic properties, while fracture and damage evolution were addressed only using macro phenomenological models. Hence, pursuing an advanced micromechanical model to predict the local and global responses, including fracture and failure envelopes, is highly beneficial.

The parametric high-fidelity generalized method of cells (PHFGMC) is a refined micromechanical computational method aimed at solving detailed and relatively large-size computational problems [1]. Several works have already used the general method of cells to capture the elastic properties of a filament wound C/C-SiC [2] and evaluate the thermo-mechanical behavior of a C/SiC CMC [3]. However, these studies have limitations, such as assumed linear displacement in the subcells (volumes), ignoring the inter-phase tension shear coupling. More importantly, the complex microstructure is idealized with simplified geometrical representation (potential inaccurate local stresses).

A unique quality of the proposed PHFGMC is its ability to derive spatial (three-dimensional) mechanical properties out of a two-dimensional repeated unit cell (RUC). For three-dimensional (3D) RUCs, the PHFGMC allows efficient calculation (using fewer degrees of freedom) relative to other numerical approaches. The current study introduces the PHFGMC to predict the mechanical behavior of C/C-SiC CMC based on computed tomography (CT) scans. Both nano-CT (having a 1-micron resolution) and micro-CT (10-micron resolution) techniques were used to extract high-fidelity multi-phase geometrical models. The proposed multiscale PHFGMC framework enables a realistic depiction of the material structure and allows for calculating the effective orthotropic properties of a single layer. The advantage of the high-fidelity geometric refined model over an idealized one is also shown by capturing local stress concentration areas related to real-depicted microstructures. The proposed modeling can be further developed to establish a reliable simulation of C/C-SiC damage-tolerance mechanical behavior.

[1] Haj-Ali R, Aboudi J. A new and general formulation of the parametric HFGMC micromechanical method for two and three-dimensional multi-phase composites. International Journal of Solids and Structures.2013;50(6):907-19
[2] Pineda EJ, Fassin M, ReeseS, Simon J-W. Method of cells-based multiscale modeling of elastic properties of filament wound C/C-SiC including free Si and matrix porosity. Journal of Materials Science & Technology.2019;35(12):2906-18
[3] Skinner T, Rai A, Chattopadhyay A. Multiscale ceramic matrix composite thermomechanical damage model with fracture mechanics and internal state variables. Composite Structures.2020;236:111847.

* Corresponding mail: rami98@tau.ac.il

AUTHOR INDEX

Faculty of Aerospace Engineering
Technion - Israel Institute of Technology
Department of Aerospace Technology
Haifa 32000
Israel

ISBN 978-1-7138-9694-4